AF500152

# Software Engineering for Manufacturing Systems

## IFIP – The International Federation for Information Processing

IFIP was founded in 1960 under the auspices of UNESCO, following the First World Computer Congress held in Paris the previous year. An umbrella organization for societies working in information processing, IFIP's aim is two-fold: to support information processing within its member countries and to encourage technology transfer to developing nations. As its mission statement clearly states,

> IFIP's mission is to be the leading, truly international, apolitical organization which encourages and assists in the development, exploitation and application of information technology for the benefit of all people.

IFIP is a non-profitmaking organization, run almost solely by 2500 volunteers. It operates through a number of technical committees, which organize events and publications. IFIP's events range from an international congress to local seminars, but the most important are:

- the IFIP World Computer Congress, held every second year;
- open conferences;
- working conferences.

The flagship event is the IFIP World Computer Congress, at which both invited and contributed papers are presented. Contributed papers are rigorously refereed and the rejection rate is high.

As with the Congress, participation in the open conferences is open to all and papers may be invited or submitted. Again, submitted papers are stringently refereed.

The working conferences are structured differently. They are usually run by a working group and attendance is small and by invitation only. Their purpose is to create an atmosphere conducive to innovation and development. Refereeing is less rigorous and papers are subjected to extensive group discussion.

Publications arising from IFIP events vary. The papers presented at the IFIP World Computer Congress and at open conferences are published as conference proceedings, while the results of the working conferences are often published as collections of selected and edited papers.

Any national society whose primary activity is in information may apply to become a full member of IFIP, although full membership is restricted to one society per country. Full members are entitled to vote at the annual General Assembly, National societies preferring a less committed involvement may apply for associate or corresponding membership. Associate members enjoy the same benefits as full members, but without voting rights. Corresponding members are not represented in IFIP bodies. Affiliated membership is open to non-national societies, and individual and honorary membership schemes are also offered.

# Software Engineering for Manufacturing Systems

## Methods and CASE tools

**IFIP TC5 international conference on Software Engineering for Manufacturing Systems, 28 – 29 March 1996, Stuttgart, Germany**

Edited by

**Alfred Storr**

*University of Stuttgart*
*Germany*

and

**Dennis Jarvis**

*CSIRO Division of Manufacturing Technology*
*Australia*

SPRINGER-SCIENCE+BUSINESS MEDIA, B.V.

First edition 1996


Originally published by Chapman & Hall in 1996
MyCopy version of the original edition 1996

DOI 10.1007/978-0-387-35060-8



A catalogue record for this book is available from the British Library

∞ Printed on permanent acid-free text paper. manufactured in accordance with ANSI/NISO Z39.48-1992 and ANSI/NISO Z39.48-1984 (Permanence of Paper).
www.springer.com/mycopy

# CONTENTS

1

# Software Engineering for Control Technology - Definitions and Requirements

*Storr, A.*
*Prof. Dr.-Ing.*
*Institut für Steuerungstechnik der Werkzeugmaschinen und Fertigungseinrichtungen, Seidenstr. 36, 70174 Stuttgart/Germany*
*Tel: 0711/121-2420 Fax: 0711/121-2408*

**Abstract**

Software for control technology is a significant part of a machine. The necessity of applying systematically procedures and methods for the development of control software is emphasised. The use of CASE tools and the specification of reusable software modules can increase the productivity. When selecting the methods and procedures, various constraints have to be considered. The selection is influenced by the hierarchical level of the control architecture, the kind of control function that is to be implemented and the required quality features. An important basic design method is the decomposition of control software according to the decomposition of hardware features in mechanical engineering. This will lead to an object-oriented structure. Additionally, using the description method of the state graph, the reuse of control software is supported.

**Keywords**

Structures and demands of control technology; software engineering procedures and methods; CASE (computer-aided software engineering) tools

# 1 INTRODUCTION

Nowadays, the term 'software crisis' is often heard. What is generally implied is that there is a shortage of software engineers. The outcome is that engineer-like systematical procedures and methods are often not applied to software development and especially not to software design. The significance of software as a time and cost factor as well as a product and manufacturing component and consequently as machine element has not yet been completely recognized.

1. Software development is differentiated from the software application cycle. Application includes also maintenance and service i. e. further development. Experience has shown that the expenditure concerning the maintenance can be higher than that concerning the software engineering.

2. The engineering cycle consists of different actions. The sequence of these actions reflects a systematical process or procedure. It is important that much time is dedicated to the "early" actions and that they are characterized by thorough and extensive analyses.

3. Furthermore, the figure illustrates that errors often occur relatively early in the process but are usually discovered late. This results in cost-intensive corrections.

Fig. 1 shows some features of software development.

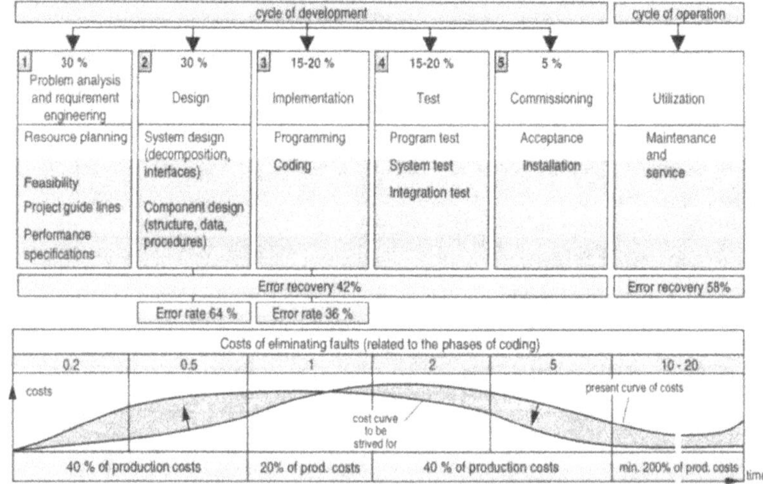


Fig. 1: Phases of software engineering, time and cost (according to Hering and others) (104 348)

The figure does not explicitly demonstrate the necessity of re-usable software and the resulting advantages as, for example, cost reduction and a reduced number of errors. Furthermore, systematical engineer-like software development requires methods e.g. for design modelling, in addition to the above mentioned process. CASE (Computer Aided Software Engineering) tools - are based on procedures and methods (fig. 2). The development and the application of function-specific CASE tools is required urgently.

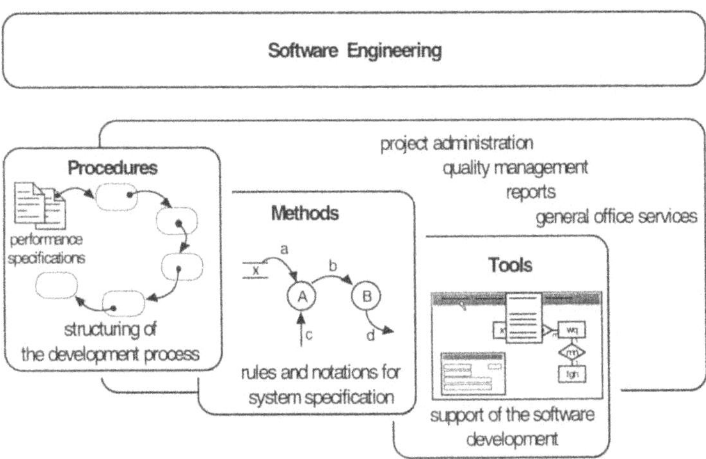


Fig. 2: Components of software engineering (104 258)

## 2 DEFINITIONS

### 2.1 Software Engineering - Software Technology

Presently both terms are used. Software engineering is according. to IEEE "the systematic approach to the development, operation, maintenance and retirement of software". According to Fig. 3, software engineering is principle-oriented, whereas software technology is understood as application-oriented software engineering. Here it is clearly seen that software technology is, to a high degree, an engineering discipline and serves for the systematical production of software. In this paper, software design in regard to the application "control technology" stands in the forefront.

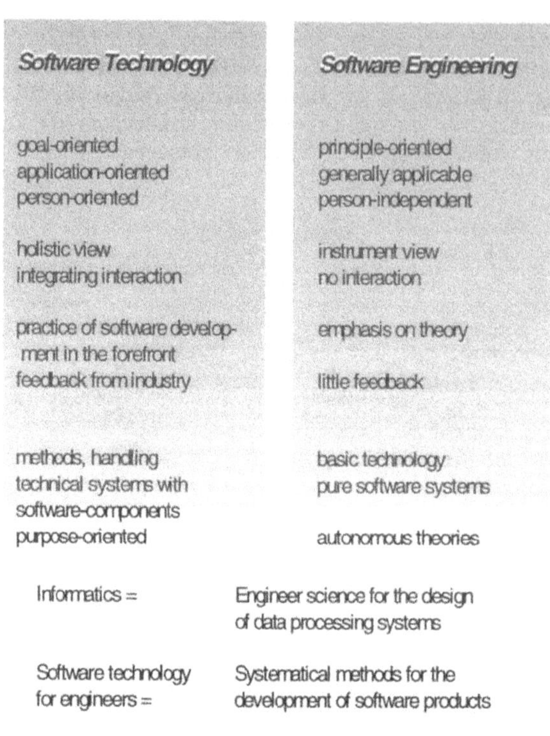


Fig. 3: Comparison software technology and software engineering (acc. to Göhner)(104 323)

## 2.2 Activities and Methods

Methods can support the execution of single or multiple tasks. When supporting multiple tasks, a method is comprehensive and continuous. Fig. 4 compares methods according to the mentioned differentiation. Furthermore, function-, data- and object-oriented methods are distinguished. It is important that a method is supported by a graphically describing or representation function. In Fig. 5 additional methods and such functions are listed. Both figures show clearly the problems with the selection of methods, especially since these do not cover the broad range of requirements. They must be assessed by their performance in regard to the application area - in our case control technology. A single method is not sufficient, but, for example, a data-oriented and a function-oriented method have to be combined.

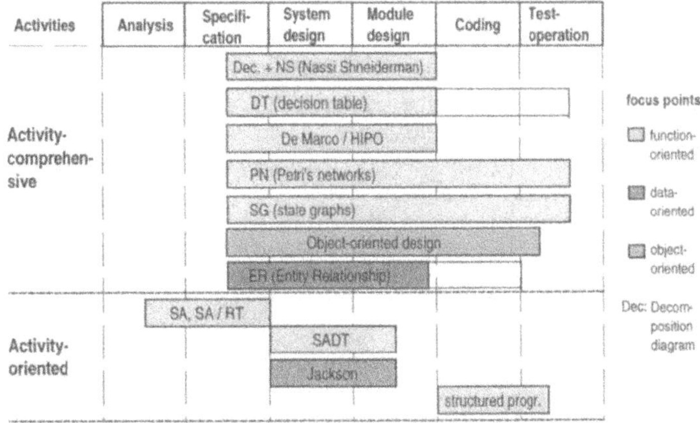


Fig. 4: Methods for software engineering (306 518)

| Description methods | Abbr. | Standards | Authors | Type FO | DO | OO |
|---|---|---|---|---|---|---|
| Decision table | ET | DIN 66241 | - | ● | - | - |
| Program flowchart | PAP | DIN 66001 | - | ● | - | - |
| Nassi-Shneiderman-structograms | NS | DIN 66261 | Nassi & Shneiderman | ● | - | - |
| Hierarchical function description | HF | - | - | ● | - | - |
| Pseudo-code | PC | - | - | ● | - | - |
| Structured Analysis | SA | - | Ross, Yourdon, Constantine, De Marco, Gane & Sarson, Ward & Mellor | ● | - | - |
| SA-Real Time | SA/RT | - | | ● | - | - |
| Structured Analysis and Design Technique | SADT | - | Ross | ● | - | - |
| Structured Design | SD | - | Yourdon, Constantine, Myers | ● | - | - |
| Specification and Description Language | SDL | - | - | ● | - | - |
| Petri's networks | PN | - | Petri | ● | - | - |
| State graphs | ZG | - | - | ● | - | ● |
| Structured control description | SR | - | - | ● | - | - |
| Entity Relationship | ER | - | Codd, Chen, Martin | - | ● | - |
| Jackson struct. programming | JSP | - | Jackson | ● | ● | - |
| Data flowchart | DFP | DIN 66001 | - | ● | ● | - |
| Design after Eiffel | | - | Eiffel | - | - | ● |
| Description after Coad, Yourdon | | - | Coad, Yourdon | - | - | ● |
| Description after Booch | | - | Booch | - | - | ● |
| Description after Shlaer, Mellor | | - | Shlaer, Mellor | - | - | ● |

FO ... function-oriented OO ... object-oriented ● yes
DO ... data-oriented - no

© ISW '96

Fig. 5: Description methods (104 214)

### 2.3 Decomposition

Decomposition - whether it be top down or bottom up - leads to modularity and hierarchy of systems. It is a basic design method and has for a long time been applied to hardware features in mechanical engineering. The objective is the definition of control functions in regard to this decomposition. This results in a communication basis for various departments of an enterprise, e.g. for mechanical and electrical design and software development. Sometimes the term machining objects is introduced. They represent an integration of a mechanical function unit and software i. e. control functions. Re-usability is the goal and thus configurability of software becomes more and more important.

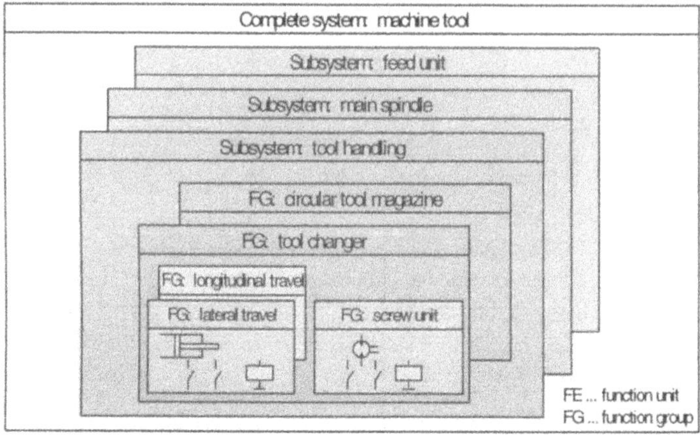


Fig. 6: Function-oriented structuring of a machine tool (301 104)

## 3 DEMANDS ON SOFTWARE DEVELOPMENT

A distinction has to be made between the application-related demands in the field "control technology" and general demands. Control technological tasks are split up in levels, according to functional aspects (Fig. 7). This does not need to correspond with the decomposition of devices; here different structures are possible.

This workshop attends to the levels E3...E5 and also to functions for the user interface. This is shown in the figure as MMI (Man Machine Interface) in several levels. Systematically generated user interfaces are based on the Seeheim model. This level structure model distinguishes between presentation, interactive control, and an application model preparing the data. The control functions of level E3...E5 contain different criteria (Fig. 8) which result in different, graphic-oriented description or modelling methods. Unified modelling causes problems. The figure illustrates the design bases related to levels and the used representation forms. Modeling or describing the control functions may be function-, data- or object-oriented, further it can be directed at static or dynamic behavior.

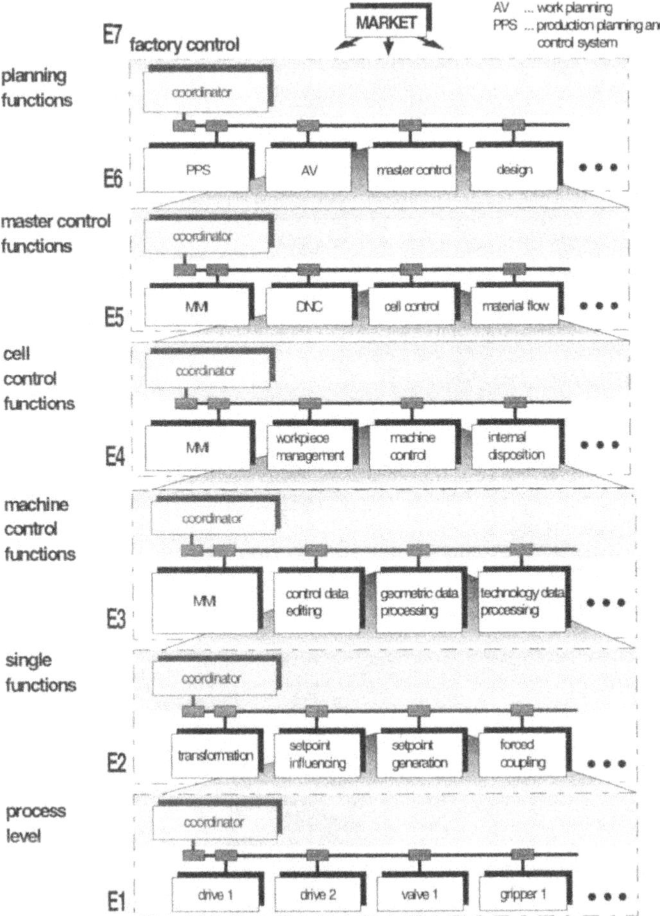


Fig. 7: Hierarchical functional control architecture (501 40/41)

Fig. 8 indicates, just like Fig. 5, that there is a problem in selecting and combining suitable methods. However, it should not be denied that different methods may contain similar elements, e.g. state transition diagrams coming from finite automatons. This has to be taken into consideration when the selection takes place.

| | Control functions/data \ Description methods | Design bases | Graphically supported representations |
|---|---|---|---|
| control functions | master control/cell level (level E5 and E4) | -state/place changes<br>-if - then - relations<br>-mathematical relations | -state/transition diagram (Z.B. PN, SG, Booch-comp.)<br>-decision tables |
| | machine control level (level E3, e.g. NC) | -mathematical relations (trigonometry, DG)<br>-closed loop control models | -block diagrams |
| | machine control level (level E3, e.g. PLC) | -logic relations (Boole)<br>-state changes<br>-sequential operations | -ladder diagram, control system function diagr.<br>-state/transition diagram<br>-sequential function charts |
| data | control levels E3 ... E5 | -data structures | -Entity-relationship-model (ERM) |

© ISW'95 DG ... differential equation SG ... state graphs PN ... Petri's networks

Fig. 8: Description methods for control function design (410 124)

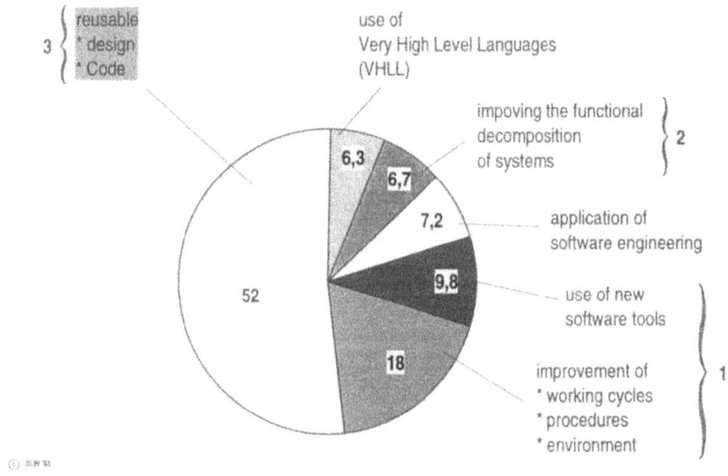


Fig. 9: Empirical analysis of the attempts made to solve the software crisis (according to Matsumoto) (104 217)

General requirements - resulting from the already mentioned problems - can be summarized in three focal points (Fig. 9):

1. Improvement and application of methods which result in so-called CASE tools;

2. Improved functional structuring of systems that is also called decomposition. In the field "control technology" this can be corresponding to the physical view at a machine and

3. Reusability of software, for example, by means of adaptability and configurability.

In fig. 10 the reusability of software is also a quality feature of software. Altogether, systematical procedures, methods and CASE tools should help to develop high-quality software. This results in a systematic software design, for which some simple examples will be presented in the following chapter.

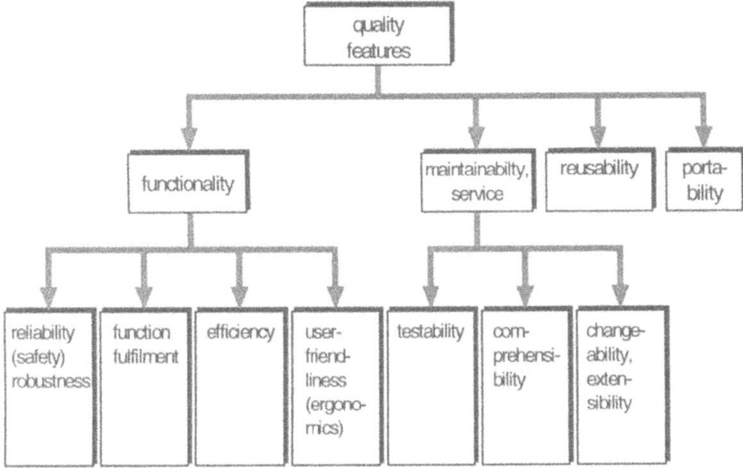


Fig. 10: Quality features of software (301 51)

## 4 CONCEPTS FOR SYSTEMATIC DESIGN

Systematic design has to comply with the generation, the re-usability of software as well as with an interdisciplinary comprehensibility of the design basis. Using the example of a feeding device (fig. 11) as can be found in a transfer line or in a transportation unit, the modelling with state graphs will be demonstrated. The decomposition is illustrated by this figure.

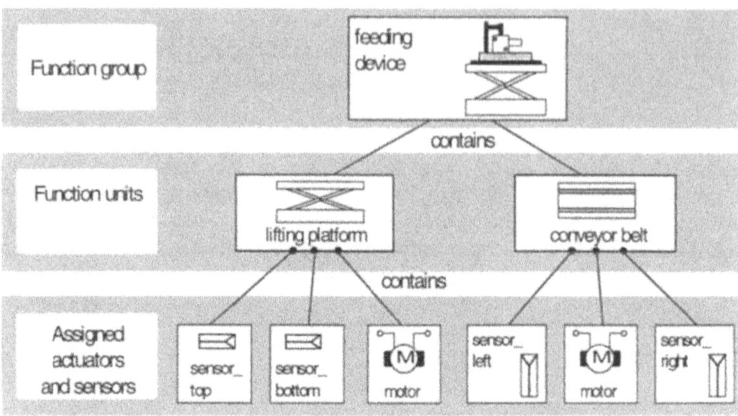


Fig. 11: Hardware decomposition of a feeding device (104 325)

The following picture illustrates the synchronisation of process sequences and function units by means of orders and acknowledgements.

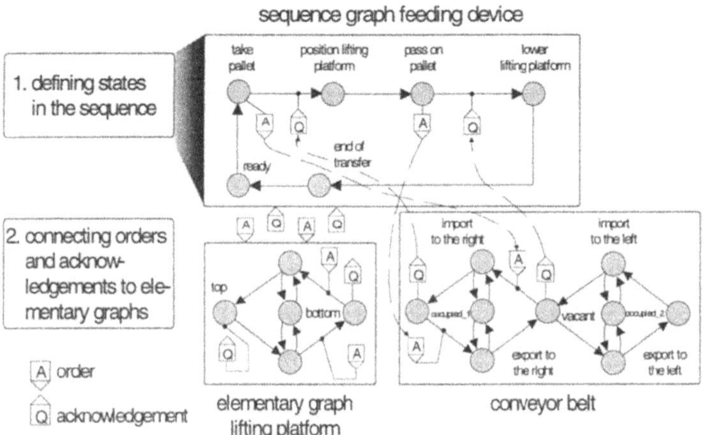


Fig. 12: Synchronisation of function group and function units (104 327)

After defining the entire description, the automized generation of the code may be done by a compiler. One example is the C code that is processed by modern controllers. Continuance up to the code is hereby achieved. Concerning the generation of the code for PLC, the conversion into the programming language of DIN IEC 1131-3 is most apt. State graphs may, as the examples proved, be called object-based because function units or sequences represent an object. Machine objects with the feature "reusability" can thus be defined.

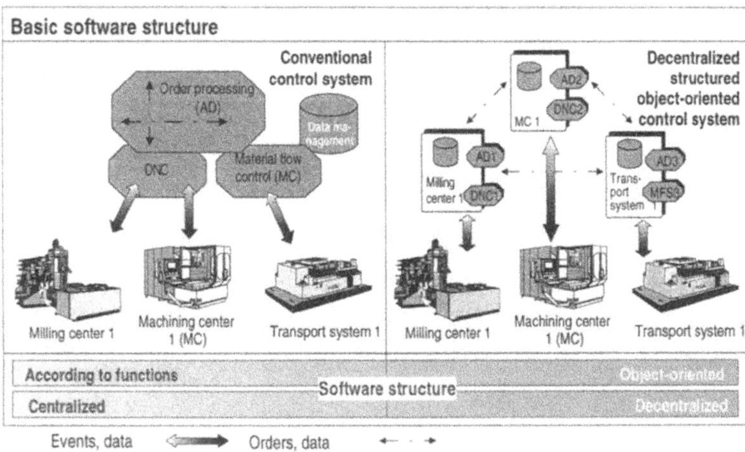


Fig. 13: Software structure of conventional and decentralized structured object-oriented production management systems (104 331)

A second example refers to the master control (fig. 7 and level E4...E5). Contemporary master controls consist of function-oriented software modules (fig. 13). The migration from a function-oriented to an object-oriented structure can be seen, with the physical object view becoming evident. The reusability and systematic design are improved considerably. Object-oriented modeling e. g. according to Booch, and programming support the systematic design by means of objects as a basis. Elements of the Booch notation are state transition diagrams related to state graphs.

Hereby the close relationship of methods for modelling various control functions of different levels becomes evident. The same applies to functions for the user interface.

## 5 SUMMARY

Concepts for computer-aided systematic design of control software and the reasons why they are so necessary were presented. Initial steps for the implementation of CASE tools were already taken in close cooperation of industry and university. The joint research work is to be continued in order to computer-aided design and generation of software by means of engineering. The demands on the different control levels must be considered as well as the computer-aided configuration of reusable modules (with classification features) and complete control systems. Furthermore, it has to be taken into account that software and information systems should contain more and more supporting elements, for example, in the field of diagnosis and if manufacturing alternatives are applicable. In addition they should be able to absorb experience and know-how of the user and to make it available if needed.

Organisational measurements for the planning of time, cost and personell with software development were not taken into account. These are necessary on top of the technical requirements mntioned in this paper, in order to execute an software development effectively.

# 6 REFERENCES


| | |
|---|---|
| Ludewig, J. | Software Engineering und CASE-Begriffserklärungen und Standortbestimmung. it 33 (1191) H. 3., S. 112 ...120. |
| Davis, A. M. | Software Requirements. New Jersey: Prentice Hall 1990. |
| DeMarco, T. | Structured Analysis and System Specification. New York: Yourdon Press 1978. |
| Ward, P.T.<br>Mellor, S.J. | Structured Development for Real-Time Systems. New York: Yourdon Press 1985. |
| Chen, P.P. | The Entitiy-Relationship Method - Towards a Unified View of Data. ACM Transactions on Database Systems, Vol. 1, No. 1 (March 1976), S. 9 ... 36. |
| Booch, G. | Booch Method of Object Analysis and Design. Santa Clara (CA): Rational 1992. |
| Rumbaugh, J. | Object-Oriented Modeling and Design. Englewood Cliffs, New Jersey: Prentice Hall 1991. |
| Pritschow, G. u.a. | Studie über die Auswirkung von einheitlichen Entwurfs- und Entwicklungswerkzeugen zur Softwareentwicklung und durchgängigen Softwaredokumentation für eine Fertigungszelle. Frankfurt: VDW-Forsch.ber. 1012, 1991. |
| Pritschow, G. u.a. | Pflichtenheft und Bewertung von CASE-Tools zur durchgängigen Software-Erstellung und Software-Doku-mentation. Frankfurt: VDW-Forsch.ber. 1013, 1993. |
| Storr, A. u.a. | Simultan zur SPS-Software - Neue Ansätze zur effizienten SPS-Programmierung. ELEKTRONIK 43 (1994) 23, S. 124 ... 136. |
| Fleckenstein, J. | Zustandsgraphen für SPS - Grafikunterstützte Programmierung und steuerungsunabhängige Darstellung. ISW 63. Berlin: Springer 1987. |

Brantner, K. Adaptierbares Leitsteuerungssystem für flexible Produktionssysteme. ISW 96. Berlin: Springer 1993.

Siewert, U. Systematische Erstellung adaptierbarer Leitsteuerungssoftware am Beispiel der Durchsetzungsplanung. ISW 100. Berlin: Springer 1994.

Pritschow, G. Siewert, U. Offene Systeme für die Automatisierung in der Produktion. GMA (VDI/VDE-Gesellschaft Meß- und Automatisierungstechnik) - Kongreß'93 Automatisierungstechnik. Dresden, 20.9.1993.

Storr, A. Uhl, J. Objektorientierte Leittechnik: neue Perspektiven und Lösungen. CIM-Management 11 (1995) 1, S. 30 ... 34.

## 7 BIOGRAPHY

Prof. Storr is the deputy director of the Institute of Control Technology for Machine Tools and Manufacturing Units at the University of Stuttgart. His main fields in research and development work are master control systems, information systems and NC programming systems, furthermore the automation of the technical information flow. These fields are accompanied by systematic software engineering.

# Life Cycle Support for PLC Controlled Manufacturing Systems.

*Jacqueline Jarvis*[a] *and Dennis Jarvis*[b]
[a]*University of South Australia, School of Computer and Information Science, The Levels, Sth. Aust., 5095, Australia. jacquie.jarvis@ unisa.edu.au*
[b]*CSIRO Division of Manufacturing Technology, P.O. Box 4, Woodville Sth. Aust., 5011, Australia. dhj@adl.dmt.csiro.au*

**Abstract**
Software support for PLC controlled manufacturing systems has focused primarily on the design and production phases of the control system lifecycle - there is limited support for the lifecycle of the manufacturing system itself. Also those phases of the system lifecycle after system development are poorly supported because effective support tools for these phases need to address both the control system and the manufacturing system. We believe that if a model of the total system is constructed at an early stage in the lifecycle which captures both the behaviour of the control system and the behaviour of the manufacturing system then the opportunity exists to develop a suite of model-based tools to support the lifecycle of a particular system from design onwards. We have tested this hypothesis by developing a methodology to enable models of PLC controlled manufacturing systems to be generated efficiently and constructing model-based software to assist in maintenance, diagnosis and system migration activities. We have also identified other activities within the lifecycle that would benefit from such software. The software that we have developed is in support of a 700 i/o point, 3 station assembly line.

**Keywords**
programmable logic controllers, maintenance, model-based diagnosis

# 1 INTRODUCTION

In certain manufacturing sectors, such as the automobile and food processing industries, the quest for automation and flexibility has resulted in extremely complex manufacturing systems, which are typically controlled by Programmable Logic Controllers (PLCs). These systems exhibit a lifecycle similar to that of conventional software systems, which involve the following phases (Birrell and Ould, 1985):

- project inception
- system definition
- system design
- system production
- system acceptance
- maintenance
- obsolescence

Manufacturing systems differ from conventional software systems in that there are two sub-systems to consider - the control system and the system being controlled (such as an assembly line). The linkage between the two systems is provided by wiring diagrams which are maintained separately from the PLC program and the pneumatic and hydraulic circuit diagrams which typically constitute the documentation for the system being controlled. This separation makes system maintenance particularly difficult because in order to understand the behaviour of the system one needs to understand how both sub-systems behave and how they interact with each other. This problem is further compounded because the information that one has access to is often inconsistent and incomplete. Inconsistencies often arise because the subsystems and their linkages are maintained using separate systems. Incompleteness occurs because typically not all the information required to understand the behaviour of the system (such as the state behaviours and the initial states of entities in the manufacturing system) is explicitly documented.

Our primary motivation was to develop an effective diagnostic system for an existing PLC controlled assembly line. Diagnostic systems for PLC controlled manufacturing systems have been developed using rule-based and other approaches (Myers and Davis, 1990; Milne et al, 1994; Wheeler and Rosetti, 1993; Day and Rostosky, 1994). However, their deployment in systems of the complexity found in the automotive and food processing industries has been extremely limited, as conventional rule-based approaches are unable to produce comprehensive systems in reasonable time-frames (Cirocco et al, 1995). The underlying reason for this is that rule-based diagnostic systems are constructed by associating a set of states with a set of observed faults. This association is normally performed by people who have had extensive experience with the underlying manufacturing system. Unfortunately, when dealing with complex, one-off manufacturing systems, it becomes impossible for a single person (or even a group of people) to experience a sufficiently wide range of fault states to generate a comprehensive diagnostic system. Furthermore, the knowledge acquisition task involved in the development of such a system is extremely time consuming (and therefore expensive), thus providing more disincentive.

It is our belief that a model-based approach offers a much more attractive path for the development of comprehensive and timely diagnostic systems. In a model-based approach, one would first construct a model of the manufacturing system. Reasoning strategies would then be developed that enabled an association to be made dynamically between an observed fault state and an underlying cause. As noted above, such a model would need to capture the behaviours of both the control system and the system being controlled. Furthermore, we expect that such models would prove to be extremely useful in other maintenance activities and in other phases of the system lifecycle. We are unaware of any attempts to construct models of this kind, although there has been some interest in representing existing control logic with alternative representations in order to facilitate analysis of the control program. Representations that have been used include AND / OR graphs (Cirocco et al, 1995; Day and Rostosky, 1994), Boolean equations (Moon, 1994; Falcione and Krogh, 1993; Boullart, 1992) and logic networks (Asfahl and Balagamwala, 1990). Wheeler and Rossetti (1993) constructed a semantic network representation of a PLC-controlled spinning line in a nylon yarn factory for diagnostic purposes. The semantic network acted as a repository for functional, behavioural, structural and heuristic knowledge relating to the line. Of particular interest to us is that the behavioural knowledge was generated automatically from the PLC programs. However, the outcome of this process was not a dynamic model that could be executed, but rather a static representation with equipment states represented explicitly as nodes in the semantic network.

The purpose of this paper is to demonstrate that behavioural models that encompass both the control and manufacturing aspects of a PLC controlled manufacturing system can be effectively generated and that such models can then be used to develop applications which are of use in various stages of the system lifecycle. Our initial objective was to support activities in the maintenance phase (in particular fault diagnosis), but we have also developed applications which could support activities in the system design and system acceptance phases.

## 2 MODEL CONSTRUCTION

### 2.1 The Pilot System

The pilot system was a low-volume assembly line. It consists of 3 assembly stations (stations 10, 20 & 30) linked by a transfer line. Two types of product are assembled on the line - Style A and Style B. The operations performed at each station are an alternating sequence of automatic and manual steps. Automatic operation at a station is initiated by an operator (or operators) depressing one or more sets of palm buttons. Those buttons remain depressed for the duration of that step. The activities performed during automatic operation typically involve the opening and closing of clamps. On completion of an automatic step (indicated by a lamp being illuminated on the station control panel), the operator removes his or her hands from the palm buttons and initiates a sequence of manual activities. This may involve the fixing of components to the sub-assembly, the loading of components, or perhaps removal of the sub-assembly from the station using external lifting equipment.

When the assembly operations have been completed at a station, the operator needs to wait until assembly has been completed in the other stations. At that point, the transfer line is activated (by the operators at each station simultaneously depressing their palm buttons) and the cycle begins again with a new sub-assembly. Note that the number of stages in a cycle depends on the product being built.

## 2.2 The Model

A model of a PLC controlled manufacturing system, if it is to be used for fault diagnosis, will need to simulate the following activities:

*a. PLC operation*
A PLC executes the algorithm described in Figure 1:

```
for (ever)
{
  Read all inputs;
  Evaluate the PLC program;
  Set all outputs;
}
```

**Figure 1** A simplification of the control loop typical of a PLC.

One loop is known as a *scan*. Our objective was not to emulate the detailed operation of the PLC so that accurate scan times can be determined, but rather to simulate its input / output behaviour. That is, having read all the PLC inputs, we need to determine what PLC outputs should be set by evaluating a representation of the PLC program.
*b. Manufacturing System Operation*
The PLC controls a manufacturing system. We can view the manufacturing system as consisting of 2 different entities:
1. Agents
2. Sensors

An agent is defined as a collection of electromechanical devices that can exist in one of several states. State selection is controlled by one or more PLC outputs. One or more events in the manufacturing process are associated with each agent state. As an example, agent SAV23 in the pilot system consists of two solenoids, an air valve, a piston and 3 clamps. The agent has two states designated SAV23A and SAV23B. In the first state, the clamps are open; in the second state, the clamps are closed. Sensors enable us to determine whether a particular manufacturing event has occurred. Examples of sensors include limit switches, proximity switches and palm buttons. These entities typically interact as shown in Figure 2:

```
Set a PLC output;
The agent changes state and associated manufacturing events occur;
                                              /* eg clamp 123 closes */
if (there is a sensor for this event / state)
{
   The new state is detected;            /* eg by limit switch 456 */
   The associated PLC input is set;
}
```

**Figure 2** The sequence of events initiated by the setting of a PLC output.

Our primary interest is the causal relationships that exist between agent states and sensors. We are not concerned with the detailed operation of the individual agents but rather their state behaviour, as can be seen from the simulation trace illustrated in Figure 3:

```
*************************************************************
      Stage 1 Ready. Hands On - Clamp Body
*************************************************************


      Y527 = 1 -> <2,1>
      Previous State
      ST20 7A-A,7B-B LOCATOR UNCLAMP BOTH SIDES
      X346 X345 X348 X347 X438 X437 -> OFF

      Current State
      ST20 7A-A,7B-B LOCATOR CLAMP BOTH SIDES
      X403 X436 X435 -> ON

      Y529 = 1 -> <2,1>
      Previous State
      ST20 7D-D,7F-F,7J-J LOCATOR UNCLAMP BOTH SIDES
      X351 X350 X360 X359 X353 X352 -> OFF

      Current State
      ST20 7D-D,7F-F,7J-J LOCATOR CLAMP BOTH SIDES
      X404 -> ON
```

```
Y548 = 1 -> <1,2>
Previous State
ST20 OVERHEAD FRAME CLAMP
X482 X481 X486 X485 -> OFF

Current State
ST20 OVERHEAD FRAME UNCLAMP
X484 X483 X488 X487 -> ON
```

**Figure 3** Simulation trace for part of stage 1 for station 20.

Agents and sensors were implemented as finite state machines; agents change states when the appropriate PLC outputs are fired. When an agent undergoes a state transition, sensors associated with the previous state are deactivated, and those for the current state are activated. The PLC program was represented as a sequence of AND / OR graphs (one per rung); rung evaluation was viewed as a single-source shortest path problem for a graph with unweighted edges[1] (Weiss, 1993).

# 3 APPLICATIONS

Our primary objective has been to develop applications in support of the maintenance phase of the lifecycle presented in Section 1. However, in developing these applications, we have identified aspects of the system design and system acceptance phases which would benefit from model based tools. In a related activity (Jarvis and Jarvis, 1996) we have investigated the feasibility of transforming PLC controlled manufacturing systems into distributed systems consisting of intelligent agents known as holons. One can view this as either a design or maintenance activity; we choose to view it as a maintenance activity.

## 3.1 Maintenance

Within the maintenance phase, we have developed tools to assist in system validation, fault diagnosis and system redesign. We also believe that documentation based on a system model would offer considerable advantages over existing approaches.

*a. System Validation.*

For each station, model validation proceeded in two phases. The outcome of Phase 1 was a model which, when started from the initial standby state, achieved cycle completion. Furthermore, the model could be run for more than one cycle (for this to happen, the state on cycle

1. Our objective is to determine whether a suitable path between the start node and the finish node exists, not to calculate the shortest such path. Consequently, we can use a shortest path algorithm, but stop as soon as a suitable path is found.

completion needs to be the standby state). In Phase 2, a PLC memory image was captured at completion of each stage of the station cycle and compared with the state predicted by the simulation. Much to our surprise, not only did these comparisons highlight further modifications that needed to be made to our model, but they also identified several limit switches which had failed in such a way that the station was still completing its cycle (and consequently maintenance personnel were unaware of the problems). This was possible because of the way in which the system was designed. If one inspects the pneumatic circuit diagrams, one will see that in many cases, only one agent state (e.g. clamp open) is explicitly checked for with a limit switch. If the PLC program then only checks to see whether that event has happened, and if the limit switch associated with that event fails in the *on* position, PLC execution will continue as normal and the station will cycle normally. (Note however, that if the clamp fails to operate correctly for some reason, we have the potential for major damage to occur). A system for the detection of such faults (which we called benign) was developed and installed.

*b. Fault Diagnosis*

In constructing our diagnosis system, we assumed that

- there will be a single point of failure
- the point of failure will be a sensor (typically a limit switch)

The justification for the first assumption is that independent multiple faults which all contribute to a line stoppage are unlikely to occur. The second assumption can be relaxed if required; if it proves necessary we can induce faults in PLC outputs[1]. Note that this would allow us to diagnose multiple *dependent* faults; e.g. if a cylinder failed to advance, then all limit switches associated with all clamp actions associated with cylinder advancement would be in error.

The conceptual strategy that we used for diagnosis is presented in Figure 4.

```
Capture the PLC state when an automatic step fails to complete
Determine the stage during which failure occurred
for (all possible faults in the station)
{
  simulate the behaviour of the station with that fault induced
  if (the simulation fails in the same stage as the physical system)
  {
    compare observed state and simulated state
    if (observed state == simulated state)
       report cause of fault and exit
  }
}
report failure to identify cause of fault
```

**Figure 4** Conceptual strategy for model-based fault diagnosis.

1. This was not done in the current version because maintenance personnel found fault-finding in this situation relatively straightforward. Their major difficulty is in finding faults associated with a single PLC input.

In implementing this strategy, there are two major issues that need to be considered:

1. How do we determine the fault space?
2. How do we compare observed and simulated states?

The fault space for a particular station is the set of all possible faults that can occur. As we are dealing with a system which changes with time, a faulty limit switch will be characterised not only by its state (*on* or *off*) but also by the point in the cycle when the failure occurred. At first glance, the size of the fault space[1] would suggest that the diagnostic strategy described above would be infeasible. However, as one would expect, the fault space can be pruned significantly. (Alternatively, degenerate fault states could be removed by inductive learning (Bratko et al, 1989)).

Comparison of observed and simulated fault states is not as complicated. The issue here is "What is the minimum amount of information that will enable us to identify a simulated state as being the same as the observed fault state?" We have chosen in the current implementation to do the comparison on the basis of selected inputs[2]. This has worked well in practice, although in general, a PLC state will be characterised by all its inputs, latched outputs and non-volatile C memory locations.

The diagnostic system was validated by inducing faults during normal production and then attempting to ascertain the cause of the resultant stoppage by using the diagnostic system. The success rate during the validation and commissioning stages was ~95%. It takes 20-40 seconds to perform a diagnosis; the system is currently running on a dedicated Sun IPC configured with 8Mb of RAM.

*c. System Redesign*

While PLCs have played a major role in the implementation of cost-effective automation, the style of programming that is required is not conducive to the development of modular control systems. It is worth noting that complex manufacturing systems (such as assembly stations) have a well defined hierarchical structure. However, it is difficult to construct PLC based control systems that reflect that structure. If one could do this, we believe that the verification and maintenance tasks for such systems would be significantly reduced. This problem is recognised by industry, and one solution that is emerging is to use separate PLCs for separate assembly functions. However, this solution is only being used for new systems and the problem of maintaining the integrity of the total manufacturing system still remains.

The concept of a holonic manufacturing system has recently emerged as a realistic candidate for the creation of the flexible and modular systems that the manufacturing sector will increasingly require (Deen, 1993; McFarlane et al, 1995). Holonic manufacturing systems will be constructed from entities known as holons, which will exhibit the dual characteristics of autonomous behaviour and the ability to function cooperatively. As such, they are similar to the multi-agent systems of distributed AI (Jennings, 1995). While holonic architectures offer the potential for systems which exhibit improved flexibility, reconfigurability and fault toler-

1. For Station 20, there are approximately 17,000 points in the full fault space. This is the product of the number of input points (210 for Station 20) and the number of scans in the cycle (75 for Style A, 88 for Style B).

2. Due to production pressures, maintenance personnel were unable to prepare a system which was free of benign faults. Consequently, we chose to ignore those inputs which corresponded to benign faults.

ance, most manufacturers are unlikely to embrace this new technology if its adoption requires them to discard their existing manufacturing and control systems. Manufacturers have an enormous investment in existing control technology, and what is required is a methodology which enables existing systems to be progressively evolved into holonic systems. We have demonstrated that the first step in this process, namely the creation of skeletal holonic subsystems which co-exist with conventional control systems is indeed feasible and that software can be developed which will enable the holonic subsystems to be automatically generated from a model of the existing manufacturing system (Jarvis and Jarvis, 1996). Note that we assume that holons have already been identified and correspond to the agents and sensors in our existing model.

In the holonic model, the only entities that will appear are holons. The control function is achieved by each holon maintaining knowledge of the states that it can adopt (e.g. clamp open, clamp closed) and the preconditions for its state transitions to occur. Preconditions are specified in terms of holon states. In the current model, notification of state transitions is achieved by a broadcast mechanism - when a holon changes state, all holons are notified. The key issue in the conversion of the PLC-based model to a holonic model thus becomes the extraction of preconditions for state transitions from the PLC-based model. The approach that we adopted (Jarvis and Jarvis, 1996) involved the following steps:

- transformation of the PLC-based model into an equivalent model with PLC inputs replaced by agent states.
- generation of a complete simulation trace for the agent state model
- extraction of the preconditions for every state transition from the simulation trace
- input the preconditions generated in the previous step into the holonic model and execute the model.

The approach has been tested successfully on the pilot system. An agent state model for the pilot system was generated, and preconditions were extracted for all state transitions. A holonic model of a subsystem of one of the stations was then constructed using the IDPS distributed programming environment[1].

*d. Documentation and User Training.*

The model has been used by people unfamiliar with the the assembly line to gain an understanding of its operation. The usage of this capability has been limited because of the absence of an appropriate user interface. If an interface was provided which enabled maintenance engineers to interact with the model in an intuitive way (e.g. point and click on a ladder logic diagram), users could interact with the system in a "what if?" mode. Thus, user-directed diagnosis could be undertaken by inducing faults in the model and observing consequences. This mode of interaction would also be extremely useful in enabling engineers to gain an understanding of complex systems and to assess the impact of proposed changes to either the control system or the manufacturing system. Note that if our model was augmented with timing information for

1. IDPS is a programming environment which supports the development of prototype IDPS-OS applications. IDPS operates in a UNIX / TCP/IP environment. IDPS-OS (Seki et al, 1991) is a distributed, fault tolerant operating system which was implemented using reliable broadcasting between objects.

the manufacturing events, the opportunity exists to explore the impact of control system modifications on cycle time.

As discussed earlier, the construction of a complete manufacturing model requires information to be extracted from several different sources. A consequence of this is that it is difficult to ensure that the information contained in all the sources is correct and consistent. Furthermore, we have highlighted through this exercise that the information which is normally maintained is insufficient to construct an operational model of a manufacturing system. It is therefore attractive to contemplate using the manufacturing model as the primary form of documentation. In this scenario, all information is held online and printed material (such as wiring diagrams) are generated from the online database. Consequently, changes can be better controlled, as we can now enforce a single entry point for system modifications.

## 3.2 System Acceptance

The system that we developed to detect benign faults, whilst deployed as a maintenance tool, could be deployed during system acceptance to ensure that the actual behaviour of the system corresponded to the "predicted" behaviour of the system.

## 3.3 System Design

During the course of the development and commissioning of the preceding applications, we developed a number of tools to identify potential design limitations in the manufacturing system and control system (e.g. limit switches which are only checked in one state, clamp positions which are not sensed, latches that don't latch, dependencies which could invalidate concurrent activities etc.). We recognise that if formal modelling techniques were used, such as Petri nets (Desrochers and Al-Jaar, 1995), process algebras (Milner, 1989) and temporal logics (Clarke et al, 1986), we could perform much more detailed and rigorous analyses of the system behaviour.

Petri nets have long been advocated as an alternative to ladder logic for the description of logic control systems. However, despite their advantages in terms of analysis and ability to represent concurrency, industrial perception was that they were too difficult to use (Desrochers and Al-Jaar, 1995). This led in 1977 to the development of GRAFCET, which is closely related to a subset of Petri nets called condition / event nets. However, no analysis can be done using GRAFCET, and place- and transition-invariants cannot be derived, which are essential for the verification of models. Also constraints on GRAFCET's evolution rules reduce its modelling power and applicability in applications which exhibit conflict, concurrency and asynchronous operation. Design frameworks based on Petri nets continue to be proposed (Ferrarini and Maffezzoni, 1993) but they have not as yet gained widespread acceptance. Frameworks based on finite state automata are beginning to emerge (Brandin and Charbonnier, 1994). This approach has the advantage that the concept of finite state machines is one which is well understood by maintenance engineers. With both approaches, a major challenge is to develop frameworks which can coexist with existing technology. In the short term, that may mean having design tools which generate PLC code.

The extent to which formal techniques can be used for the analysis of existing PLC controlled manufacturing systems is an issue that needs further investigation. In particular, the ability to effectively generate alternative representations for existing systems needs to be addressed. We have demonstrated through the creation of our holonic models that it is feasible to transform the existing system representation into a finite-state representation. Other transformations (to temporal logic (Moon, 1994) and to Sequential Function Charts (Falcione and Krogh, 1993)) have been reported. In both cases, the starting point is a user-generated control specification consisting of Boolean equations. Our experience with the maintenance procedures for large, complex systems suggests that if we want such files to accurately represent the control program, they must be generated automatically from the control program. Consequently , a key component of our modelling methodology is software that parses ladder logic listing files and produces AND / OR graphs. This software could be easily extended to generate Boolean equations and other representations as required.

Once an alternative representation has been constructed, one can then use analysis tools appropriate for that representation. A problem here is that for large systems, the state space is potentially enormous and analysis may become infeasible. A solution to this problem is to create models with higher levels of abstraction. Coloured Petri nets (Jensen, 1992) provide such a capability, as does the Circal process algebra (McCaskill and Milne, 1992). Unfortunately PLC programs do not lend themselves to abstraction because they indicate a functional, rather than a structural decomposition of a system. A much better starting point if abstraction is required is provided by the holonic models described above.

## 4 CONCLUSION

We have developed a model of an existing PLC-controlled assembly line which incorporates the behaviour of both the manufacturing system and the control system. This model was then used to develop a collection of model-based applications to support the maintenance and redesign of the assembly line. In the course of that work, we developed tools which would be of use in the design and acceptance phases of PLC-controlled manufacturing systems. We have also identified other applications which would benefit from the availability of behavioural models or from formal models.

## REFERENCES

Asfahl, R.C. and Balagamwala, A. (1990), Simlog: a Prolog Based Simulator for Industrial Logic Control Systems. *Computers and Industrial Engineering*, **19**, pp. 195-199.

Birrell, N.D. and Ould, M.A. (1994), *A Practical Handbook for Software Development*, Cambridge University Press.

Brandin, B. and Charbonnier, F. (1994), The Supervisory Control of the Automated Manufacturing System of the AIP, In *Proc. of the 4th. International Conference on Computer Integrated Manufacturing and Automation Technology,* Troy New York, pp. 319-324.

Boullart, L. (1992), Using AI Formalisms in Programmable Logic Controllers. In Boullart, L., Krijgsman, A. and Vingerhoeds, R.A. (Eds.) *Application of Artificial Intelligence in Process Control*, Pergamon Press, pp. 96-113.

Clarke, E.M., Emerson, E.A. and Sistla, A.P. (1986), Automatic Verification of Finite-State Concurrent Systems Using Temporal Logic Specifications. *ACM Transactions on Programming Languages and Systems*, **8**, pp. 244-263.

Cirocco, L.R., Jarvis, D.H., Jarvis, J.H. and Ryan A. (1995), Simulation of a PLC Controlled Assembly Station. Technical Report MTA 333, CSIRO Division of Manufacturing Technology, Adelaide, South Australia .

Day, W.B. and Rostosky, M.J. (1994), Diagnostic Expert Systems for PLC Controlled Manufacturing Equipment. *International Journal of CIM*, **7**, pp. 116-122.

Deen, S.M. (1993), Cooperation Issues in Holonic Manufacturing Systems. In Yoshikawa, H. and Goosenaerts, J. (Eds.) *Information Infrastructure Systems for Manufacturing (B-14)*, Elsevier Science, pp. 401-412.

Desrochers, A.A. and Al-Jaar, R.Y. (1995), *Applications of Petri Nets in Manufacturing Systems: Modeling, Control and Performance Analysis.* IEEE Press.

Falcione, A. and Krogh, B. (1993), Design Recovery for Relay Ladder Logic. *IEEE Control Systems*, **13**, pp. 90-98.

Ferrarini, L. and Maffezzoni, C. (1993), Conceptual Framework for the Design of Logic Control. *Intelligent Systems Engineering*, **2**, pp. 246-256

Jarvis, J.H. and Jarvis, D.H. (1996), A Strategy for Migration of a PLC-Controlled Manufacturing System to a Holonic Manufacturing System. In preparation.

Jennings, N.R. (1995), Controlling Cooperative Problem Solving in Industrial Multi-Agent Systems Using Joint Intentions. *Artificial Intelligence*, **75**, pp. 195-240.

Jensen, K. (1992), *Coloured Petri Nets. Volume 1: Basic Concepts*. Springer-Verlag.

McFarlane, D., Marett, B., Elsley, G. & Jarvis, D. (1995), Application of Holonic Methodologies to Problem Diagnosis in a Steel Rod Mill. In *Proc. of the 25th. IEEE Conf. on Systems, Man and Cybernetics,* Vancouver, pp. 940-945.

McCaskill, G.A. and MIlne, G.J. (1992), Hardware Description and Verification Using the Circal System, Research Report HDV-24-92, University of Strathclyde, Department of Computer Science, Glasgow, Scotland.

Milne, R., Nicol, C., Ghallab, L., Trave-Massuyes, L., Bousson, K., Dousson, C., Quevedo, J., Aguilar, J. and Guasch, A. (1994), TIGER: Real-Time Situation Assessment of Dynamic Systems, *Intelligent Systems Engineering*, **3**, pp. 103-124.

Milner, R. (1989), *Communication and Concurrency,* Prentice Hall International Series in Computer Science.

Moon, I. (1994), Modeling Programmable Logic Controllers for Logic Verification. *IEEE Control Systems*, **14**, pp. 53-59.

Myers, D.R., Davis, J.F. and Hurley, C.E. (1990), A Knowledge Based Approach to Malfunction Diagnosis of Discrete Operations Involving Programmable Logic Controllers. In *Proc. of the 1990 American Control Conference*, San Diego, pp. 1974-1979.

Seki, T., Hasegawa, T., Okataku, Y. and Tamura, S. (1991), An Operating System for the Intellectual Distributed Processing System - An Object Oriented Approach Based on Broadcast Communication, *Journal of Information Processing*, **14**, pp. 405-413.

Weiss, M.A. (1993), *Data Structures and Algorithm Analysis in C.* Benjamin / Cummings.

Wheeler, G. and Rossetti, V. (1993), Model Based Diagnosis in a PLC controlled Factory. In *Proc. of Object Oriented Software*, London.

## BIOGRAPHY

Jacqueline Jarvis is a Lecturer in the School of Computer and Information Science at the Universty of South Australia. She holds a B.Sc. degree from the Flinders University of South Australia, and an MSc. in Software Development and Analysis from Heriot-Watt University. Her current research interest is model-based fault diagnosis.

Dennis Jarvis is a Principal Research Scientist at the CSIRO Division of Manufacturing Technology in Adelaide, South Australia. He holds a B.Sc. degree from the Flinders University of South Australia, and a Dip. Comp. Sci. and a Ph.D. degree from the University of Queensland. His current research interests are in cellular manufacturing, enterprise modelling and model-based fault diagnosis. He is a member of AAAI.

# State Diagrams A New Programming Method for Programmable Logic Controllers

*Dipl.-Inf. Hans-Peter Otto*
*Siemens AG*
*Nuremberg-Moorenbrunn, Germany, Phone 0911/895-2612, Fax 0911/895-2122*

*Dipl.-Ing. Günter Rath*
*Siemens AG*
*Nuremberg-Moorenbrunn, Germany, Phone 0911/895-2253, Fax 0911/895-2122*

**Abstract**

A new fast and reliable way of writing programs for PLCs is the state diagram programming method. This method enables both sequential and non-sequential processes to be described in a graphical form. Automation tasks are broken down into function units, of which the states and their coordination are described graphically by state diagrams. Each state diagram has only one active state, but can exchange information with other diagrams for coordination purposes. The result is a set of interlinked state "automatons". Grouping, for example by machine subassemblies, gives the automation task a clear structure.

The subject of applying the state diagram concept to PLC programming was also dealt with by a working committee of the VDW (Association of German Machine Tool Manufacturers) to increase the efficiency of PLC programming. On the basis of these results a programming tool executable under Windows 95 has been developed which enables state diagrams to be used for programming PLCs right through from the draft stage to testing, documentation and diagnostics.

**Keywords**

State Diagrams, Messages, Templates, Views, Status

# 1 METHOD

A new fast and reliable way of writing programs for PLCs is the state diagram programming method. This method enables both sequential and - especially - non-sequential, asynchronous processes to be described in a graphical form. The key advantage for the user is that this method of representation is not only suitable for PLC programmers but also for mechanical engineers, commissioning engineers and service engineers.

The application of the state diagram concept to PLC programming with a view to increasing efficiency in PLC programming was also dealt with by a working committee of the VDW (Association of German Machine Tool Manufacturers) / 4 /. The Automation Group of SIEMENS AG has used these results to develop HiGraph, a tool for the new SIMATIC S7 PLC family which enables you to use state diagrams for programming PLCs right through from the design stage to testing, documentation and diagnostics.

Before programming automation tasks with state diagrams, you must break them down into (mechanical) function units, e.g. the actuators of a machine. You then describe the action of each function unit using state diagrams. The diagrams show **states** in the form of circles and **transitions** in the form of arrows.

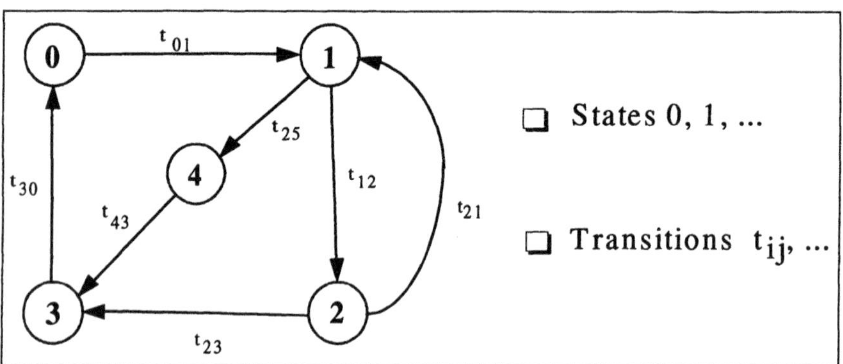


Fig. 1: Elements of a state diagram

States can describe either static "processes" (e.g. rear stop position) or dynamic processes (e.g. forward motion). The transitions represent the permissible transitions between states. States can also be used to trigger actions and transitions can be assigned conditions.

You can use messages to "synchronize" the function units and/or "coordinate" them with a higher-level state diagram. In this way you can obtain a set of interlinked state automatons.

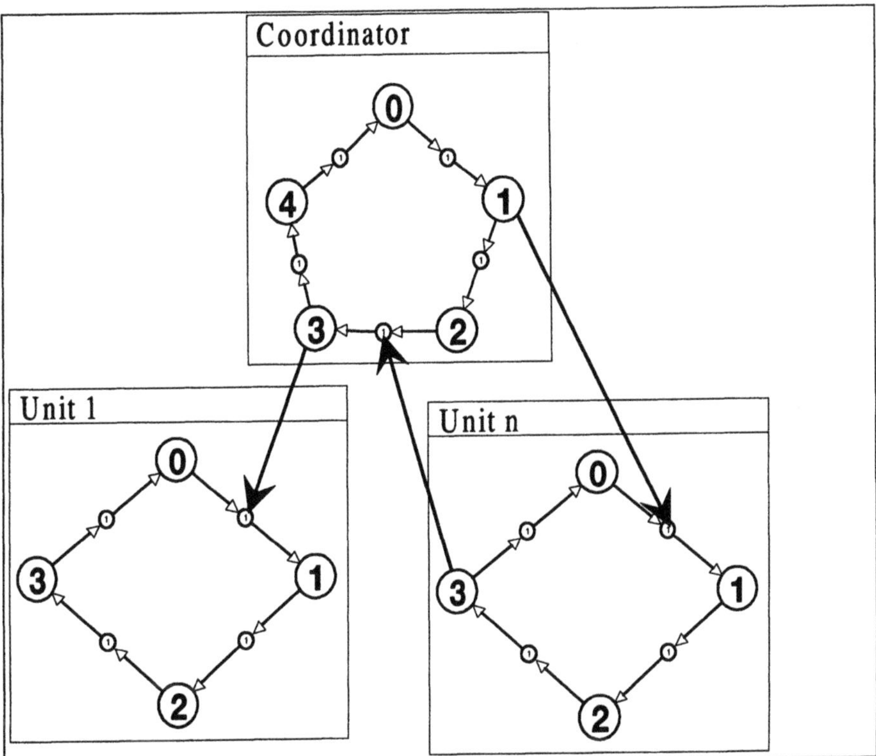


Fig. 2: Exchange of messages between diagrams

It is often helpful to divide the function units into groups (e.g. to correspond with the subassemblies of a machine) and use a coordinator for each group. This gives the PLC program a clearer structure and improves the reusability of the individual sections of the program.

# 2 PROCEDURE FOR PROGRAMMING A PLC USING STATE DIAGRAMS

## 2.1 Task definition

Before beginning to solve an automation problem a **definition** of the task must be available. This is normally provided by a **specialist engineer**. First of all, we expect to receive a **process schematic** containing all the main components (function units) of the machine, subassembly or plant to be controlled and clearly showing the mutual effect of their actions. As an example, the diagram below shows the rotary table of a milling machine for the machine tool industry with the following function units:

- Motor
- Index
- End support (fixing clamp)

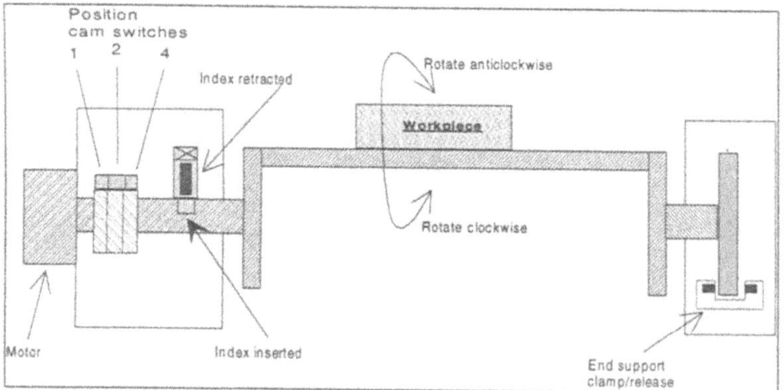


Fig. 3: Example of a rotary table

In addition to the process schematic the programmer also expects to receive a **functional description**. In machine building this usually consists of a function chart. But what is to prevent the engineer from creating this using a state diagram editor instead of drawing it on paper? In our example, the chart would look like this:

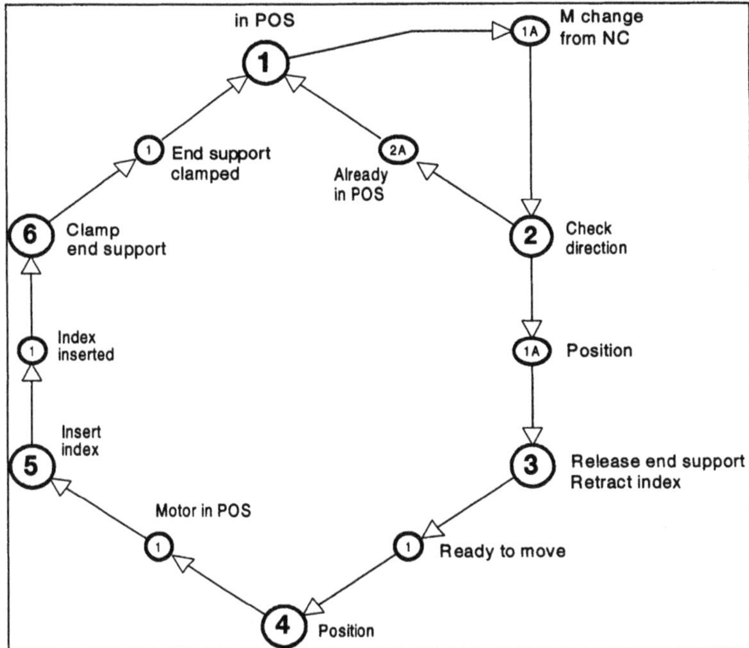


Fig. 4: Coordination chart for rotary table

Using the HiGraph editor also has the advantage that the resulting diagram can be passed on to the programmer for expansion and the addition of finer details.

## 2.2 Programming

*Defining the objects in the process*

A task definition as described above comprises both *process schematics* and *functional descriptions* in the form of state diagrams. The process schematics must now be checked to ensure that they contain all the function units (motors, valves, etc.). To check the functional descriptions it is advisable to divide the machine into subassemblies such as spindle, revolver, etc. with their respective function units. You then assign a state diagram to each function unit. Since the functions of the individual subassemblies can be performed independently of one another, it is necessary or at least desirable to provide each subassembly with an additional state diagram as a coordinator.

The rotary table in the task definition above, for example, consists of three function units: the motor, the index and the end support (fixing clamp). A state diagram is provided for each of the function units and one for coordination (i.e. 4 diagrams in our example).

*Describing functions rather than "programming" them*

The "programmer" now uses state diagrams to **describe** each of the function units and the coordinator, i.e. to define the possible states and the transitions between these states. It is convenient to proceed step by step as follows:

- Describe normal operation (how you envisage the function)
- Describe initialization, i.e. the conditions to be observed when switching on the controller
- Analyze possible abort criteria for motion states, e.g. Reset
- Define the conditions for exiting abort states
- Define the monitoring precautions (e.g. monitoring times for motion states).

You must also define the messages for coordinating the state diagrams.

*Task-oriented diagram views*

If all the operating states of the function units are included, the diagrams tend to become rather unwieldy. The graphic for a state diagram is therefore divided into a number of levels, which can be displayed as required. These levels are called views because they each display a different aspect of the state diagram.

The display reproduced below shows a diagram in Basic/Normal view.

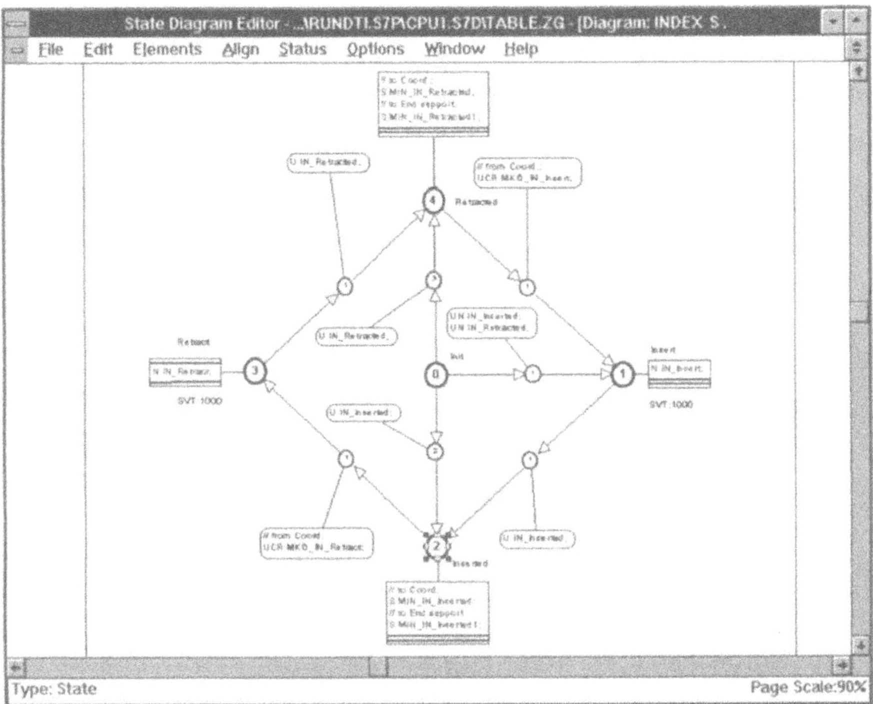


Fig. 5: State Diagram in "Normal" View

*Wait and monitoring times*

HiGraph allows you to activate a wait time and a timeout for each state.

A **wait time** causes the diagram to remain in the state in which the time was activated for at least the set time.

A **timeout** enables you to monitor the timing of motion states. If there is no state change within the timeout set, an error message is issued containing the diagram group number, the diagram number and the number of the state in which the error has occurred. In this way powerful error diagnostics can easily be integrated at the drive and coordination level.

*State-independent functions*

It is frequently necessary to keep a constant check on certain criteria (monitoring or abort conditions) independently of the current state in order to change to the relevant error or abort state. This is the purpose of the **Any transition,** which allows changes from any state of the diagram to a defined state. This enables limit switches to be monitored for non-permissible dual use, for example. If the target state is designated as an error state, an error message is automatically issued.

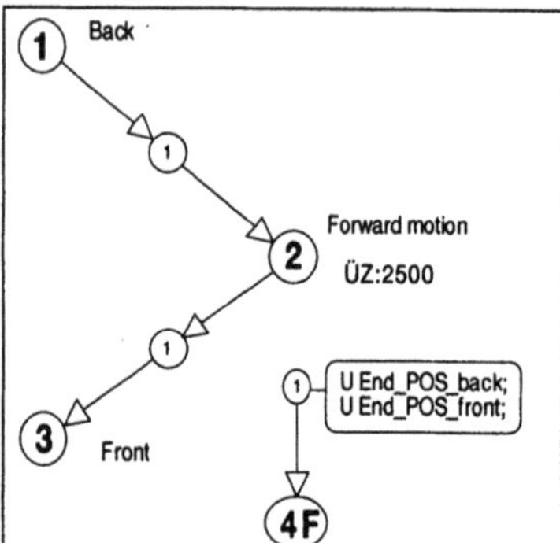


Fig. 6: Diagram elements for error diagnostics

*Reusing diagrams*

Considerable rationalization can be achieved by reusing software which has already been tested. In automation systems, this is particularly effective with the function units at drive level. For example, you can generate and test the functions for valves with one, two or three stop positions or for motors as part of a new project and use them again afterwards. For maximum flexibility of reuse, HiGraph provides **templates.** For this purpose, all signals (inputs, outputs and messages) are treated as formal parameters to be defined in the interface declaration for each diagram. When using a template you merely assign the addresses of the signals to the formal parameters. If you want to store the diagram as a template, you store the entire diagram with the exception of the addresses assigned in the interface declaration. By entering the addresses of the signals of a different function unit you can reuse the complete functionality. If modifications are necessary for the new functional unit, you can make them without affecting other function units.

## 2.3 Program testing

Once the program is in the PLC, the commissioning engineer can conveniently start to test it using the same method of representation as that used by the engineer and programmer to describe the tasks. The **Status** display at graphical level, which marks both the current state and the last transition, makes the functions more transparent. The graphic is constantly updated to keep pace with the process, so that the user can see the current transition and the subsequent state of the state diagram. This means that errors in the program logic can readily be detected by people who were not involved in the programming. The following section of a display illustrates the Status function.

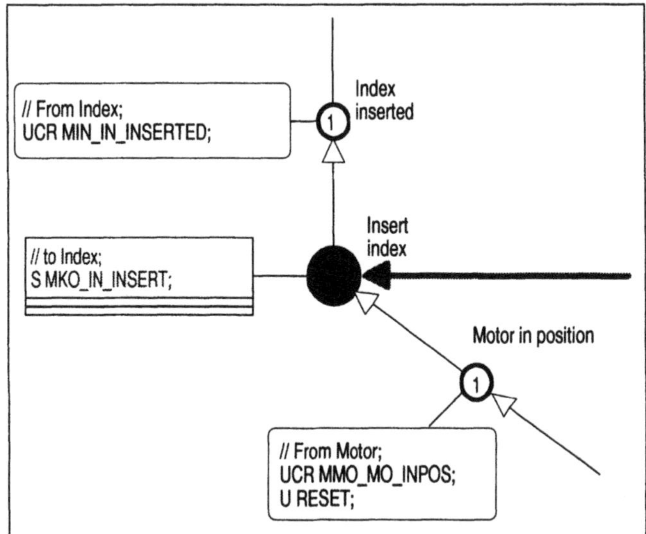


Fig. 7: Section of display showing status

If changes need to be made to a function in the program, they can be made in the graphic display or in the zoom, as when writing the program. The program is then recompiled and transferred to the PLC again. Since this can be done while the PLC is in operation, the program test can be continued immediately.

## 2.4 Service and diagnostics

Service is required when the plant or machine is out of operation as a result of a fault. In the majority of cases, the fault is not in the PLC and it can be used for diagnosis. HiGraph provides support in the form of monitoring times and Any transitions, as previously described. This enables faults to be located and corrected quickly and reliably, thus minimizing machine downtimes.

On operator panels with alphanumeric display facilities a diagnostics file can be displayed with HiGraph. This enables you to use the texts entered at the programming stage for a plaintext display of the error location and an indication of the status of the signals affected.

## 3 SUMMARY

The following features distinguish the state diagram programming method from the PLC programming methods commonly used at present:

- **Object-oriented**
  For each function unit (of a machine) you create a state diagram, which gives a graphical description of all the possible states of the unit. The mode of presentation is very easy to understand.
- **Reusable**
  You can store the state diagrams as templates. By adapting the formal parameters for the inputs and outputs of other function units the diagrams can be used again for the same functions. Adjustments can be made for special attendant circumstances without affecting other function units.
- **Clear and easy to understand**
  The state diagram method is the first method to provide a clear graphical representation of the sequence **and** unit control or drive level. You can label and enter comments for every state and every transition.
- **Process independent**
  Whether you are dealing with a lathe, a honing machine, a transfer line or any other control problem, you can always use the same programming tool.

HiGraph is a tool which enables you to describe, test and correct the functions of machines and plant sections with the same method of representation right through from the definition of the task by the engineer to programming, commissioning and ultimately servicing. And because the documentation automatically always includes the graphics and all the details, undocumented function changes are a thing of the past.

## 4 REFERENCES

1 H. König Beitrag zur Strukturanalyse und zum Entwurf von Steuerungen für Fertigungseinrichtungen

1 A. Herrscher Flexible Fertigungssysteme - Entwurf und Realisierung prozeßnaher Steuerungsfunktionen, Springer-Verlag

2 J. Fleckenstein Zustandsgraphen für SPS - Grafikunterstützte Programmierung und steuerungsunabhängige Darstellung, Springer-Verlag

3 VDW 1013 Pflichtenheft und Bewertung von CASE-Tools zur durchgängigen Software-Erstellung und Software-Dokumentation

## 5 BIOGRAPHY

Hans-Peter Otto studied electrical engineering and information technology (software engineering). Since joining Siemens AG, he has worked in various development and marketing departments as a systems programmer and is now section head. As Chairman of the German standardization group in DIN DKE (German Electrotechnical Commision of DIN) and as German delegate to the corresponding international IEC Working Group, he plays an important role in the international standardization of PLC technology (IEC 1131).

Günter Rath studied general electrical engineering. Since joining Siemens AG, he has worked in the field of development and marketing for programmable logic controllers and numerical controls. At present, his work focuses on the development and market entry of a PLC programming system for state diagrams.

4

# Applying simulation modeling techniques for the design and assessment of control software

*Il. Astinov, N. Todorov*
*Laboratory 'Simulation Modeling in Industry' (SMI)*
*MTF, TMMM, Technical University - Sofia (VMEI), 1756 Sofia, BULGARIA, Tel: (+359 2) 636 3784, FAX: (+359 2) 683 478,*
*E-mail: ila@vmei.acad.bg*

**Abstract**

The paper presents a non-traditional application of simulation modeling techniques as a decision aiding tool in determining the effectiveness of a control software algorithm in manufacturing systems. Normally software engineering methods and techniques would bring the development of a control software module to a level where certain predesigned functionality is achieved. In many cases however, such functionally will not guarantee a reliable operation of the overall system. Additionally, stochastic factors such as the randomness of workpiece arrival intervals, processing times, breakdowns, which influence the operation of the real system are not taken into account. An application in using simulation modeling techniques to evaluate the effectiveness of different control algorithm designs and structures, prior to their implementation in real systems is presented in this paper.

**Keywords**

Modeling, simulation, decision making, control technology

## 1. INTRODUCTION

Control software plays a significant role in the effectiveness of modern manufacturing systems. Typically control software is designed to run on general purpose or specialized computers. It would operate on the principles outlined below.

- Analog or digital signal will be fed to the computer running the control software. These signals contain information reporting the state of the system being controlled.

- The control software will process the incoming information and according to the control algorithm will react (if required) with appropriate feedback information.
- The feedback information is translated by appropriate hardware to follow-up actions on behalf of the system.

Figure 1 is an illustration of these principles which shows the information streams coming from the system. These streams being processed by the control algorithm and then relevant control information flowing back to the system as a product of the control algorithm.

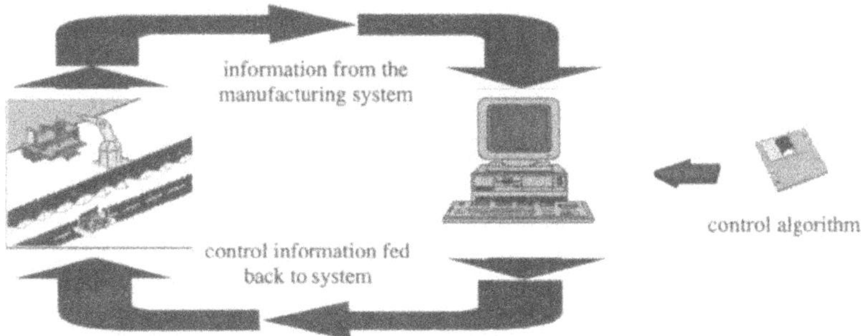


**Figure 1** Loops of information streams in manufacturing systems.

A well designed and error free code would reduce time for manufacturing, increase the system reliability yet contribute to a lower cost of the end product. The paper presents the experience of the authors in utilizing simulation modeling techniques aiming to develop effective new or improve existing control algorithms.

## 2. PROBLEM DEFINITION AND SOLUTION PROPOSED

### 2.1 Problem definition

A traditional approach in the design and implementation of such software is to utilize software engineering methodologies in developing the algorithm and code. Later on the control software is tested on the real system i.e. on the manufacturing cell, on the AGV or in the warehouse (see Figure 1). Apart from the general advantage that the performance of the real system is being observed, this approach has the following disadvantages.

- Tests generally are limited in time as manufacturing systems are to be put in operation within a strict deadline.
- Experimenting with the real system is obviously an expensive and in many cases could be a dangerous exercise.
- The majority of the stochastic factors like the randomness of workpiece arrival intervals, processing times, breakdowns, which influence the operation of the overall system can rarely be taken into account.

## 2.2 Solution proposed

In order to address extent the mentioned disadvantages, a solution to the problem was adopted, having the outline given below.

- Create a model of the manufacturing system using appropriate and familiar simulation software. This could be any general purpose software like SLAMSYSTEM or a simulator like FACTOR/AIM.
- Design alternatives to the control algorithm for the components of the system that are to be controlled. If necessary, produce code for each alternative in a language, that can be linked to the simulation software. Most simulation systems have an open architecture structure allowing user inserts, written in high level languages, such as FORTRAN 77 or C++ to be linked to the main executable file, gain access to the data structures and manipulate them.
- Create scenarios each of which will utilize a version of the control algorithm. Define clear performance measures of the system. Define and perform experiments with the scenarios of the model.
- Compare the values of the performance measures estimated as results of the simulation runs and take appropriate decisions for the choice of the most effective version of the control algorithm.

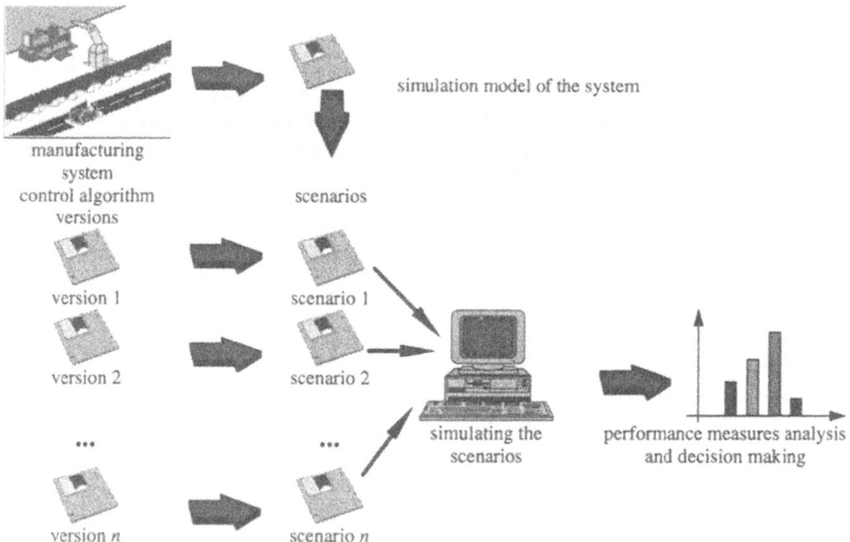


**Figure 2** Concept of the solution proposed.

Figure 2 illustrates the concept of the solution proposed. By this means simulation methodology and more particular discrete event simulation is utilized to a full extent as a powerful tool in modeling manufacturing systems (Astinov 1990, Pritsker 1986). **Yet in this particular case (Figure 2) the effectiveness of the control algorithm over the identified**

**performance measures of the system is being studied, rather than the system as a whole.** The following section describes a case study, illustrating the effect of the presented approach.

# 3. CASE STUDY - THE KILN FACILITY

## 3.1 Components and their purpose in the facility

The case study relates to a kiln facility, typical for companies manufacturing ceramic products such as tableware (cups, plates), bathroom and toilet equipment. An example of the configuration of such facility is given in Figure 3.

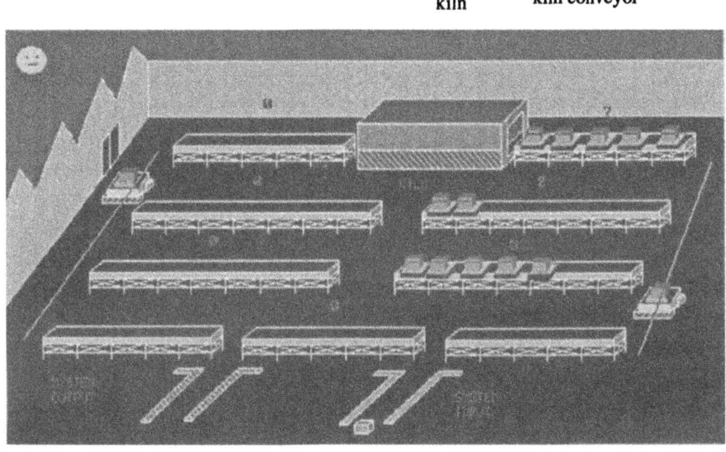


**Figure 3** Typical configuration of a kiln facility.

Items to be fired in the kiln arrive at random intervals in the input of the system. They are placed on pallets in the pallet loading station and afterwards the loading AGV would distribute them over the input conveyors. The distribution pattern **depends on the control algorithm** incorporated in the AGV controller. Once placed on the conveyor that feeds the kiln, the items are transported at a rate, that will allow them to be fired for the appropriate length of time. The conveyors after the kiln are used to hold the pallets with fired items during the night shift, as the kiln operates on a 24 hour basis, and during the night pallets with fired items are not removed from the system.

## 3.2 Versions of the AGV control algorithm

The present case study concerns an existing kiln facility, which had a particular control algorithm for the loading AGV implemented. From an operational point of view, the algorithm performed the control functions exactly as it was designed. The principle structure of the algorithm is outlined in Figure 4 (a).

**pallet with item available**
check for free space on any conveyor:
  *free space available*
    **put pallet in a uniformly chosen free conveyor space**
  *free space not available*
    **pallet waits**

(a) Outline of the existing control algorithm.

**pallet with item available**
check for free space on kiln conveyor:
  *free space on kiln conveyor available*
    **put pallet on kiln conveyor**
  *free space on kiln conveyor not available*
    check for free space on any other conveyor:
      *free space available*
        **put pallet in a uniformly chosen free conveyor space**
      *free space not available*
        **pallet waits**

(b) Outline of the proposed improvement of the control algorithm.

**Figure 4** Structures of existing and proposed AGV control algorithms.

It was thougt that the utilization of the kiln could be improved by giving priority to the kiln conveyor when a pallet with an item arrives. The outline of the proposed improvement is given on Figure 4 (b). The decision whether such a modification would be worthwhile was difficult to make because of the facts given below.

- The kiln facility as a system had stochastic properties, as the arrivals of parts at the system's input were in random intervals of time.
- The speed pallets were transported through the kiln was so low that the impact on giving priority to the kiln conveyor (whether positive or negative) was not very clear.
- If such change in the control algorithm proved to be ineffective this could lead to significant losses in terms of production quantities and energy. Such risk was not accepted by the management of the company.

## 3.3 The solution derived

The methodology described in Section 2 was employed in order to give a justified answer to the problem. A model of the kiln facility was developed using the SLAMSYSTEM general purpose simulation software (*SLAMSYSTEM* 1993). Appropriate code was developed for both versions of the AGV control algorithm.

Face validity and expert assessment were employed to validate the operation of the model, using the code of the existing AGV control software. Two scenarios were created each using one of the versions of the control algorithm. For both scenarios computer animation was produced, illustrating the dynamic behavior of the kiln facility. Separate frames of this output are given in Figure 5.

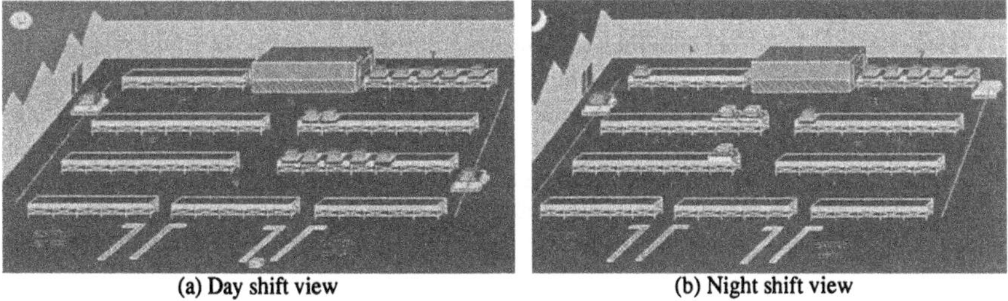


**Figure 5** Computer animation frames illustrating the dynamic system behavior.

The performance measure identified was the system throughput for 24 hours on the assumption that no other conditions of operation will be changed except the AGV control. The values of the throughput of the kiln facility estimated by the simulation runs are compared and illustrated on the bar graph given in Figure 6.

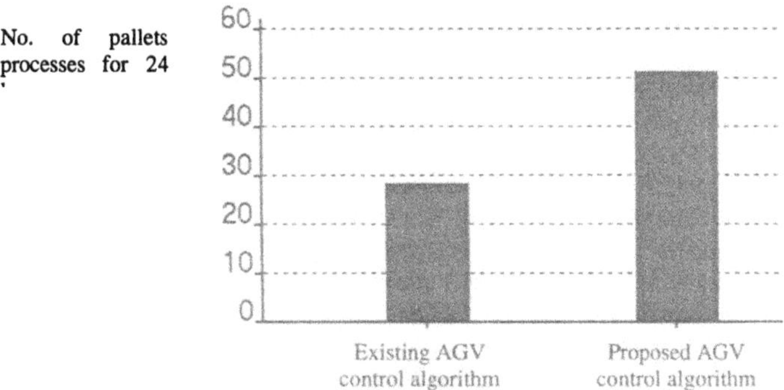


**Figure 6** Comparison of the throughput of the kiln facility utilizing different AGV control algorithms.

As it is seen from Figure 6, the implementation of the control algorithm with the structure given on Figure 4(b) will lead to an increase of the kiln facility throughput with an average of 22 pallets i.e. 39% in 24 hours. This result is a strong proof of the increased effectiveness of the new algorithm and suggests that it would be worthwhile to proceed to implementation.

This approach was extensively utilized in **two other significant projects**. The first one related to power stations using combined multiple renewable energy sources (wind, solar irradiation, batteries and diesel units). The project was sponsored by the Directorate General for Science and Technologies of the European Union under the PECO Scheme (Project ERBJOU2CT920245). Two major control approaches and related algorithms of the power station were studied. Performance measures included diesel fuel consumption, energy produced by renewables, total energy from renewables dumped in the dump load. These were evaluated through separate simulation runs. The system was strongly influenced by stochastic factors, such as power consumption, wind speed, solar irradiation and temperature. Values of each of these factors vary over time in a random manner (Astinov, 1994; Astinov Ivanova, Pavlov et al 1995).

The second project concerned the improvement of the through put of a major crossroad in Sofia, controlled by traffic lights (Astinov, Ivanova, Stoichkov et al, 1995). This project is currently under development in collaboration with the company involved in the installation and maintenance of urban traffic control systems in Sofia. The crossroad as a system is also stochastic, as car arrival intervals from each direction are random. The component in the system that can be modified is the control algorithm changing light signals for each direction. For the test case an improvement of an average of 18% was estimated for one of the control algorithms.

## 4. CONCLUSIONS

The following conclusions can be drawn, based on the results achieved so far:

- As described in Section 3.2, a version of the control algorithm may be designed according to a specified functionality. It may also be developed in appropriate code and be error free. It may operate exactly as the designers wanted it to operate. However, this is **not a guarantee** that the **control functionality incorporated** in the particular algorithm will assure **good overall system performance.** The **major advantage** of the reported approach is that it (the approach) allows the developers of control software to verify not only the proper operation and correct structure of the control algorithm, but **to evaluate the effectiveness of the algorithm over the operation of the overall system** based on redefined performance measures.
- The proposed approach eliminates the disadvantages of the development and implementation of control software given in Section 2.1. Moreover, as many versions of control algorithms as needed may be developed and studied, thus aiding the decision making process towards a rational and validated choice of suitable control algorithm functionality.
- Simulation runs can be performed for a long duration of the operation of the system, taking into account the stochastic properties of system's behavior (if available). Thus, the estimates of performance measure values will be statistically justified and correct.
- Computer animation of the system's operation is an additional tool in this approach, allowing experts to gain a realistic view of the control algorithm's influence to the overall operation of the system.

## 5. ACKNOWLEDGMENTS

The developments and results reported in this paper are part of the activities of the HC&M project ERBCIPDCT940023 sponsored under the PECO scheme of the European Union.

The activities on the simulation of combined multiple renewable systems, mentioned in the present paper were part of the activities of the ERBJOU2CT920245 project sponsored under the PECO scheme of the European Union.

Words of thank are to be said to Sonia Hristova, Hristo Dinov and Dinitar Cerov all from Traffic Signs Ltd. - Sofia for their assistance and collaboration in the crossroad project.

Finally the authors would like to express their appreciation to Zlatka Tchkarova, Wladimir Simeonov, Blagovest Stoichkov and the CMRES team - all students at the laboratory for Simulation Modeling in Industry for their valuable contributions in the mentioned projects.

## 6. REFERENCES

Astinov, Il. (1990) Simulation modeling of machine tools and systems. *PhD thesis*. Technical University Sofia, Sofia.

Astinov I.L., Bopp G., Consoli A., Lalas D.P., Morgana B., Wrixon G.T. (1994) Combined Multiple Renewable Energy Sources System Simulator Facility. *Proceedings of the European Wind Energy Association Conference and Exhibition (EWEC'94)*, Thessaloniki

Astinov Il., Ivanova D., Pavlov P, Virgili A., Leotta, Nocera V., Bopp G., Rehm M.. A.(1995); Simulating power stations driven by combined multiple renewable energy sources. *International Symposium SIELA'95*, Volume II, 183-188

Astinov Il., Ivanova D., Stoichkov B., Simeonov W. (1995) Simulation modeling in the control of traffic flow. *Proceedings of the III National Scientific Conference "Automation in Engineering and machine tool building"*, **Volume 1**, 73-76

Pritsker, A.A.B. (1986) Applications of simulation. *Introduction to simulation and SLAM II.* Halsted Press, John Wiley & Sons, New York, Chichester, Brisbane.

*SLAMSYSTEM User's Guide* (1993) Pritsker Corporation

## 7. BIOGRAPHY

**Prof. Ilario Astinov** has an MSc degree in production engineering and PhD degree in systems simulation from the Technical University - Sofia. Has over 25 publications in national and international forums. He has specialized in discrete event simulation at Purdue University (USA), Pritsker Corporation (USA), Staffordshire University (UK) and Cranfield University (UK). He is head of the laboratory for Simulation Modeling in Industry at the Technical University - Sofia and a board member of the UNIDO/DP/BUL/006/96 project. He is Deputy Coordinator of three international academic projects under the EU TEMPUS scheme and Associate Contractor to three international research projects under the EU PECO scheme involving partners from 8 EU countries.

**Prof. Nikola Todorov** has an MSc and PhD degree in the field of mechanical and production engineering. He has over 40 publications in national and international conferences and magazines. He is head of the laboratory "CAD/CAM and FMS technologies" at the Technical University - Sofia. For over 15 years he was the head of the department of CAD/CAM applications at the Institute of Machine Tools and Systems - Sofia. He is a board member of the UNIDO/DP/BUL/006/96 project.

5

# A CP-net Approach To Control Logic Engineering

*Mathew Farrington and Jonathan Billington*
*Telecommunications Systems Engineering Centre,*
*Institute for Telecommunications Research,*
*The University of South Australia.*

**Abstract**

Coloured Petri nets have been used to model and therefore specify the control logic in an automotive assembly line. This paper describes the modelling activity and shows how two models of the same system, one developed using top-down methods, the other developed using bottom-up methods, may be fused.

**Keywords**

Coloured Petri nets, Programmable Logic Controllers (PLCs), Discrete Event Manufacturing, Modelling.

## 1 INTRODUCTION

An automotive assembly line has been analysed and modelled in previous work. The models were developed using the C programming language and describe the control logic and physical operation of the line. A formal toolkit based on Petri nets is now being used to construct a model of the assembly line from the existing code. The model may then be analysed and verified ensuring that any errors in the assembly line design are uncovered and removed. The resulting CP-net model formally specifies a set of abstract system requirements against which new, distributed system models may be constructed and verified.

This paper describes the initial CP-net modelling activity in which two models were constructed, one from the top down, the other from the bottom up. Knowledge of overall system behaviour was used to construct the top-down model. At each level, agent behavioural requirements and product processing requirements were identified and modelled. The requirements placed on *agent*, *product* and *process* elements implied a more detailed, lower-level model, also consisting of agent, product and process elements. This modelling activity recursed downwards until all the behavioural and processing requirements had been resolved into their discrete control logic and machine-level events. The bottom-up model was constructed by directly representing the ladder logic and physical relationships (actuator and sensor relationships) present in the PLC/system feedback loop using CP-net structures. The final model is a fusion of these two.

# 2 TOOLS AND TECHNIQUES

Embedded manufacturing control software such as PLC logic is notoriously difficult to maintain since it represents control requirements at a very low level [Gilles, 1990]. Typically this is overcome by specifying and maintaining the code at a higher level where the control relationships are preserved in a more re-usable form [Venkatesh, *et al.*, 1994]. High level code is also more transportable because it is (in theory) platform or hardware independent and may be compiled into PLC ladder logic as required.

Even more promising though are *formal methods*. They are gaining popularity in the manufacturing control area and have the potential to achieve time and cost savings throughout the life-cycle of a system [Billington, 1991]. If a manufacturing system is designed using formal methods then specification errors are reduced (a maintenance saving); the specification may be executed to quickly gain an understanding of the system requirements (a time saving); the performance of the system may be analysed and verified; the specification may be compiled into code (thus reducing coding errors as well as lowering the *time to prototype*) - all of this *before* committing to an implementation.

The Telecommunication Systems Engineering Centre* supports a CP-net toolkit known as FORESEE [Billington, 1991]. It consists of drawing, simulation, formal analysis and implementation tools. The toolkit includes: Design/CPN†, a graphical editing, simulation and occurrence graph analysis tool; TORAS, a tool for the exhaustive simulation of Petri net specifications using state space reduction techniques [Lester, 1994]; and PROMPT a tool for the automatic conversion of Petri nets into executable code (C).

# 3 AN INTRODUCTION TO CP-NETS

## 3.1 Assembly Line Overview

A brief introduction to the assembly line is needed before the Coloured Petri net models are introduced. The automotive assembly line is divided into stages or work cells each with its own set of actuators, sensors, welding stations, conveyors etc. Each stage assembles components or attaches components to a vehicle as it progresses through the line. One PLC controls the entire system. The system is only *partially* automated. Workers must perform positioning, loading and spot welding tasks. Worker safety is an integral feature of the system control logic. For example, a conveyor will advance only when both workers in a two person work cell are at their control consoles.

A sub-section of the assembly line, known as the roofing bay, accepts a bare chassis, attaches reinforcing struts and roof panels and then releases the chassis to the next bay. The roofing bay is the focus of this paper.

---

*The TSEC web page, http://www.itr.unisa.edu.au/tsec/ .

†Design/CPN is available from Aarhus University, http://www.daimi.aau.dk/designCPN/ .

```
color Roof = with medium | long | very_long;
color Chassis = with luxury | sedan | wagon;
color Strut = with regular_strut | wagon_strut;
color Welder = with welder;

color Workers = with worker;

color WorkerAndInfo = product Workers*Chassis;
color WorkerAndStrut = product Workers*Strut;
color WorkerAndRoof = product Workers*Roof;

color ChassisAndStrut = product Chassis*Strut;
color Car = product Chassis*Strut*Roof;

color Bay = union BayChassis:Chassis +
                  BayChassisStrut:ChassisAndStrut +
                  BayCar:Car;

fun strut_type(c:Chassis):Strut = case c of
                                      luxury => regular_strut
                                    | sedan => regular_strut
                                    | wagon => wagon_strut;

fun roof_type(c:Chassis):Roof = case c of
                                    luxury => long
                                  | sedan => medium
                                  | wagon => very_long;

var roof: Roof;
var chassis: Chassis;
var strut: Strut;
var car: Car;
```

**Figure 1** Global Declaration Page for the Assembly Line

## 3.2 CP-nets by Example

It is assumed that the reader is familiar with basic Petri net concepts [Reisig, 1992]. Coloured Petri nets (CP-nets) are essentially ordinary Petri nets (Place Transition nets) with typed tokens and complex net inscriptions [Jensen, 1992]. CP-nets can specify and model systems at an implementation independent level. CP-nets are based on a well defined semantics - a system specified in CP-nets may be formally analysed or verified before implementation. CP-nets can model concurrent processes and are therefore naturally suited to the design and modelling of concurrent manufacturing systems. In this short introduction to CP-nets their semantics and dynamics are demonstrated by example.

Consider Figures 1 and 2. This CP-net was constructed using Design/CPN, a tool in which types and inscriptions are written in a functional programming language called CPN ML. Types or *colour sets* may be arbitrarily complex and are defined on a declaration page associated with the CP-net. In Figure 1 the colour sets `Roof`, `Chassis`, `Strut`, `Welder` and `Workers` are enumeration types. The `WorkerAndInfo`, `WorkerAndStrut`, `WorkerAndRoof`, `ChassisAndStrut` and `Car` colour sets are comprised of tuples. `Bay` is a *union* of the colour sets `Chassis`, `ChassisAndStrut` and `Car`. `roof`, `chassis`, `strut` and `car` are typed variables. Functions can be defined as well: `fun strut_type()` takes a `chassis` as input and returns the strut that is required for that chassis. Similarly, `fun roof_type()` returns the type of roofing panel.

Consider the CP-net structure in Figure 2. A place may only contain tokens which be-

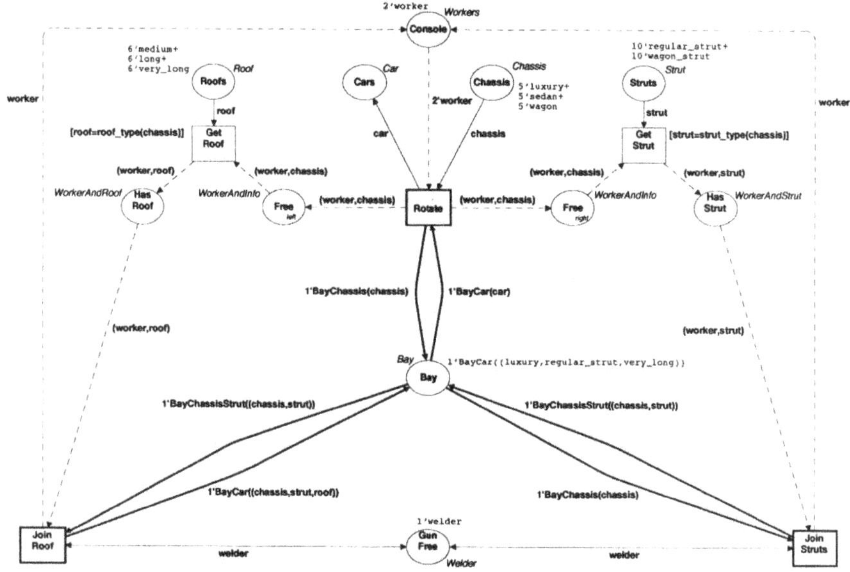


**Figure 2** Top Level CP-net of the Assembly Line

long to its colour set. For example, the place `Roofs` has a colour set `Roof` and hence may only contain the tokens `medium`, `long` and `very_long`. Adjacent to the place `Chassis` is the inscription '`5'luxury + 5'sedan + 5'wagon`'. This is an *initial marking* (of a place). The inscription is a *multi-set* where multiple instances of a particular token are represented by a preceding integer and apostrophe. Arcs may be inscribed with expressions of arbitrary complexity (including functions) that evaluate to multi-sets (over the colour set of the associated place). A *guard* is a boolean expression associated with a transition. The expression must be true for the transition to occur. The transitions `Get Roof` and `Get Strut` are guarded (by the expressions in square brackets).

Execution of a CP-net is also best demonstrated by example. Consider the transition `Rotate`. `Bay`, `Chassis` and `Console` are input places and `Bay`, `Cars` and the two `Free` places are output places with respect to `Rotate`. The three input places have non-empty initial markings. For `Rotate` to be enabled there must be sufficient tokens on the input places to satisfy the input arc expressions. This is the case: two `worker` tokens are at the `Console`; a `car` token is in the `Bay`; and there are `chassis` tokens in the `Chassis` place (five instances of three colours). Variables `car` and `chassis` must be *bound* to tokens before `Rotate` is enabled. The bindings in Figure 3 are possible. `Rotate` occurs for *one* of these bindings. Assuming that the transition occurs with binding $b_3$, the resultant marking of output places $\texttt{Free}_{left}$, `Cars`, $\texttt{Free}_{right}$ and `Bay` are given in Figure 4.

$b_1$ =<car=(luxury,regular_strut,very_long), chassis=luxury>
$b_2$ =<car=(luxury,regular_strut,very_long), chassis=sedan>
$b_3$ =<car=(luxury,regular_strut,very_long), chassis=wagon>

**Figure 3** Possible Bindings for Rotate

$M(Free_{left})$ = 1`(worker,wagon)
$M(Cars)$ = 1`(luxury,regular_strut,very_long)
$M(Free_{right})$ = 1`(worker,wagon)
$M(Bay)$ = 1`BayChassis(wagon)

**Figure 4** Rotate Occurred for $b_3$

# 4 MODELLING THE ASSEMBLY LINE

## 4.1 Top-Down Modelling

A top-down approach to modelling recognises that even the most simple models are of some benefit when it comes to understanding system behaviour and may be built upon as more detail is required.

Petri nets are well suited to the task. First a simple system is constructed with only the most basic or important characteristics and then simulated. Once satisfied with the net's behaviour, extra conditions or control structures may be added or transitions may be replaced with sub-nets until every significant element of the system is represented in the model. This technique has been well documented [Jensen, 1992, Zhou and Leu, 1991, Zhou *et al.*, 1989, Desrochers and Al-Jaar, 1995].

It is useful to introduce some terminology at this point. Manufacturing systems make *products*. The market ready product which leaves a system differs from the raw products which enter because it has been *processed* by the *agents* which make up and are *controlled* by the system. This generic terminology groups workers with agents and worker behaviour with control systems.

More precisely then: *processes* are (collections of) manufacturing events which transform products (usually from a less evolved to a more evolved state); *agents* are the entities (physical or logical) which enable and perform processes (and so transform or transport products); *control systems* coordinate the agents.

Some extra terminology is needed to support the notion that a manufacturing system consists of many smaller systems (from factory, to assembly line, to workbay, $\cdots$, to clamp). Processes may be unresolved or fully resolved. An *unresolved* process is a high-level process (such as adding a roof) where control implementation details are not provided. An unresolved process may represent many other processing steps in a manufacturing system (e.g., the top-most unresolved process in automotive assembly is *'build car'*). A *fully resolved* process is a machine-level (or worker-level) *event* which may not be resolved any further. The agents which enable these events are called *control agents* and correspond to controller or worker requirements (typically PLC output register values or worker behaviour).

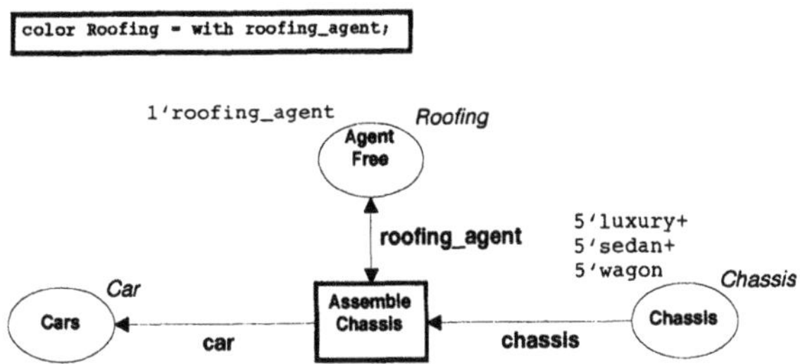


**Figure 5** The Roofing Bay as a Single Transition

## 4.2 Applying the Top-Down Approach to the Bay

A full description of the methodology is given in [Farrington and Billington, 1996]. It may be summarised as follows: Places are assigned to each agent state and product location. A transition is assigned to each process. Agent states are made input places of those processes for which they are pre-conditions and output places of those processes for which they are post-conditions. If a transition represents a fully resolved or machine level process then the agent states preceding or proceeding the transition correspond to discrete control logic or machine-level events. Tokens are defined for each agent. Agent states are represented by tuples. Tokens are defined for each product. Product locations are represented by union colour sets. Arcs are then drawn linking pre-condition places to the respective processes, and processes to the respective post-conditions. Each unresolved process must now be resolved. Local products, processes and agents are again identified and a new CP-net structure built to reflect the extra resolution in the model.

The top level CP-net of the roofing bay consists of of a single transition representing the entire roofing process and places to represent the roofing bay agent and the input and output chassis buffers (see Figure 5). CP-nets are hierarchical. The transition `Assemble Chassis` in Figure 5 represents the sub-net in Figures 1 and 2. Consider now the unresolved process `Join Roof` in Figure 2. The top-down methodology has been used to construct a CP-net for `Join Roof` (see Figure 6). The processes (transitions) in Figure 6 are *fully resolved* (i.e., they are machine-level events) and so define the control logic that must be provided to enable the correct event sequencing, the actuators and sensors needed to communicate this information and the *man machine interface* (MMI) required to integrate the automatic and manual steps in the roofing bay.

Similarly for the other unresolved processes in the assembly line model, the top-down approach yields a set of control logic requirements. These requirements may be converted into PLC logic if needed. However, in this modelling exercise the requirements are not developed any further. Instead the existing control logic (PLC ladder logic) is directly modelled and developed from the bottom up. The bottom-up models are then fused with the top-down models so that extra system details can be incorporated into the (until now idealised) model of the assembly line. The final, fused model may then be analysed, executed, verified and form the basis for future modelling activities.

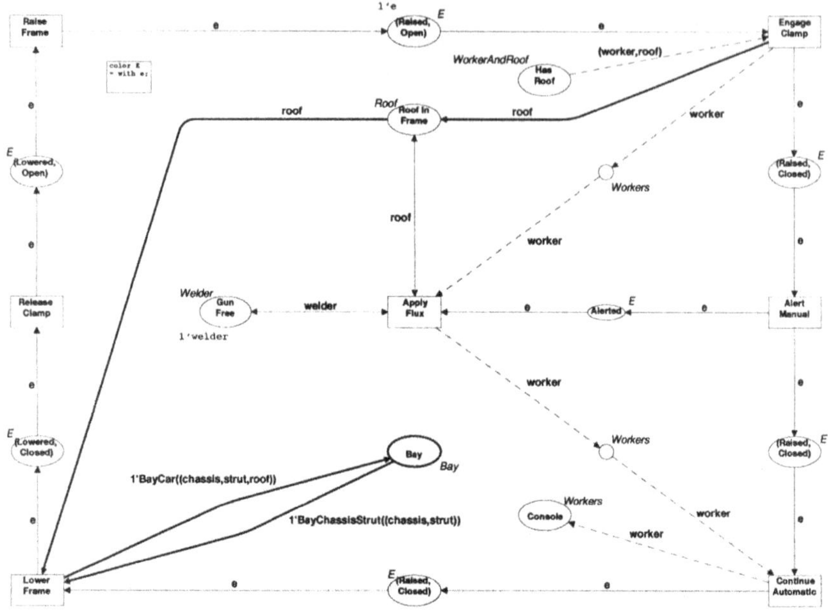


**Figure 6** The Roofing Process Resolved

## 4.3 Bottom-Up Modelling

The translation of PLC ladder logic into Petri nets is already being automated, with some researchers proposing algorithms for the design of Petri nets in particular application areas [Venkatesh *et al.*, 1994]. Such automated techniques produce ordinary Petri nets (monochromatic CP-nets). These may then be *collapsed* into much smaller CP-nets. The collapsing or colouring process recognises that large systems consist of subsystems with duplicate functionality. Large systems can therefore be modelled using a single CP-net structure and different coloured tokens for each sub-section rather than by many duplicate nets. Domain specific tools have appeared that automate this process [Darabi and Jafari, 1994].

In this paper a PLC architecture is modelled by CP-nets. Input, output and internal registers, the PLC program and, when needed, a representation of the discrete steps in a scan cycle are all included in the model. The control lines and sensor lines used to attach a PLC to the system may be described by state vectors and are modelled using lists. Hence the model is already collapsed a great deal compared with the approaches reported above.

The hardware which makes up the assembly line may be described in terms of machine-level states and events. For example, a clamp is an agent which may be *open* or *closed*. The events which place the clamp in these states are *release* and *engage* respectively. A PLC-based feedback loop similar to the one shown in Figure 7 controls the assembly line.

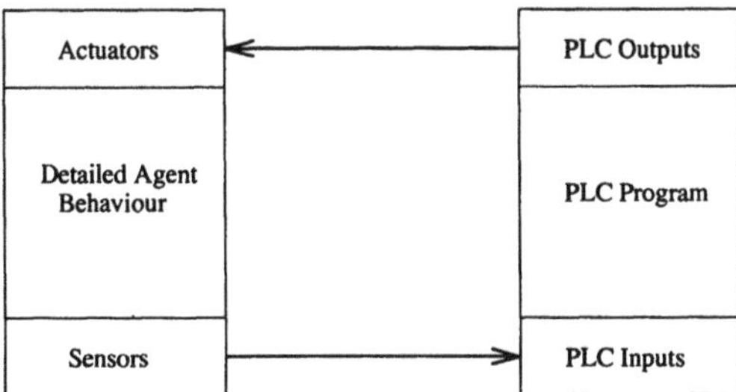


**Figure 7** The Bottom-Up Model of the Assembly Line Control Loop

Many of the agents in the assembly line are connected to sensors. Sensors simply translate the state of an agent into the on-off input signals that are recognised by the PLC. A PLC program executes repeatedly, operating on the PLC inputs and a set of variables held inside the PLC. The program calculates a set of outputs, the states of which determine the operation of actuators which update the state of the agents in the assembly line. The state of the clamp, for example, is measured by a sensor connected to a PLC input. If the clamp is open the PLC input is switched on. If the clamp is closed the same PLC input is switched off. The clamp state is dependent on the actuator which switches it between states, and so is also dependent on the PLC output which drives the actuator.

The control loop is modelled by the Figures 8 and 9. In this generic model the agent behavioural dependencies are represented by the single transition `Map Agents`. The sub-net associated with this transition defines the relationship between the elements in each agent. For example, the PLC controls the raising and lowering of the overhead frame by toggling a single switch. There is however no sensor which tests for the frame being in its correct lowered state - rather a proximity switch tests for the clamp being lowered. Hence, the `Map Agent` sub-net includes a transition which enforces the movement of the frame and clamp in unison - reflecting the fact that they are joined.

The CP-net in Figure 9 executes as follows: An initial token in the place `Sensed Agents` lists the initial states of the assembly line agents which are connected to sensors. The transition `Set Sensors` occurs, removing the token and placing a new token in `Sensor Values`. The new token (returned by the function `sensor_op()`) now represents the agent states in terms of PLC control signals. `Scan Inputs` occurs, copying this state information into the `Input Registers` of the PLC. The transition `Execute Logic` represents a sub-net which contains a CP-net model of the PLC ladder logic and performs the logic scanning function. `Execute Logic` occurs, removing the token from the place `Input Registers` and the initial token from the place `Internal Registers`. The two lists of register values are converted into a form which allows the `Execute Logic` sub-net to correctly simulate scanning behaviour. At the end of the scan the new output and internal register values are converted into lists and copied into the `Output Registers` and `Internal Registers` respectively. The transition `Set Outputs` then occurs. The output register values are converted into actuator switch settings (or switching requests) by the function `set_switches()`. The transition `Switch Agent State` then occurs, with the function `set_agent()` updating the states of those agents which have actuating circuits connected directly to the PLC outputs. The transition `Map Agents` then occurs, its

```
.
.
.
color Input = product X*X_Range*Contents;
color Internal = product C*C_Range*Contents;
color Output = product Y*Y_Range*Contents;

color Agent_State = list Agent
color Switches = list Switch;
color IP_Reg = list Input;
color Int_Reg = list Internal;
color OP_Reg = list Output;

var state: Agent_State;
var switch_vector: Switches;
var ip_reg: IP_Reg;
var int_reg: Int_Reg;
var op_reg: OP_Reg;

fun set_switches(opr:OP_Reg):Switches = ...
fun set_agent(swv:Switches):Agent_State = ...
fun sensor_op(sta:Agent_State):IP_Reg = ...

.
.
.
```

**Figure 8** Global Declaration Page Fragment for the Control Loop

sub-net updating the states of any other components in the manufacturing system. The execution cycle then repeats.

### 4.4 Fusing the two Models

Figure 10 summarises the fusion process. Every sub-net which contains machine-level processes and control agents is replaced with a new, *fused* sub-net, which more closely resembles the CP-net in Figure 9. For example, in Figure 6 the places and transitions corresponding to the overhead frame agent (such as place (`Lowered,Open`) and transition `Raise Frame`) are replaced with a new representation of the frame based on the `AGENT` CP-net in Figure 9. The places and transitions in Figure 6 which correspond to product locations or worker agents are preserved and incorporated in the new model. Each *fused* `AGENT` CP-net is then connected to the single `PLC` CP-net which controls the entire bay (and the entire assembly line).

## 5 CONCLUSIONS

Top-down modelling techniques and CP-net tools have been used to develop control logic specifications for a PLC-based automotive assembly line. Here the system was modelled in terms of products, agents and processes. Bottom-up modelling techniques and CP-net tools have been used to construct a model of the PLC logic, the physical relationships between actuators and sensors and the control loop found in the assembly line.

The initial top-down modelling activity promoted system understanding while the

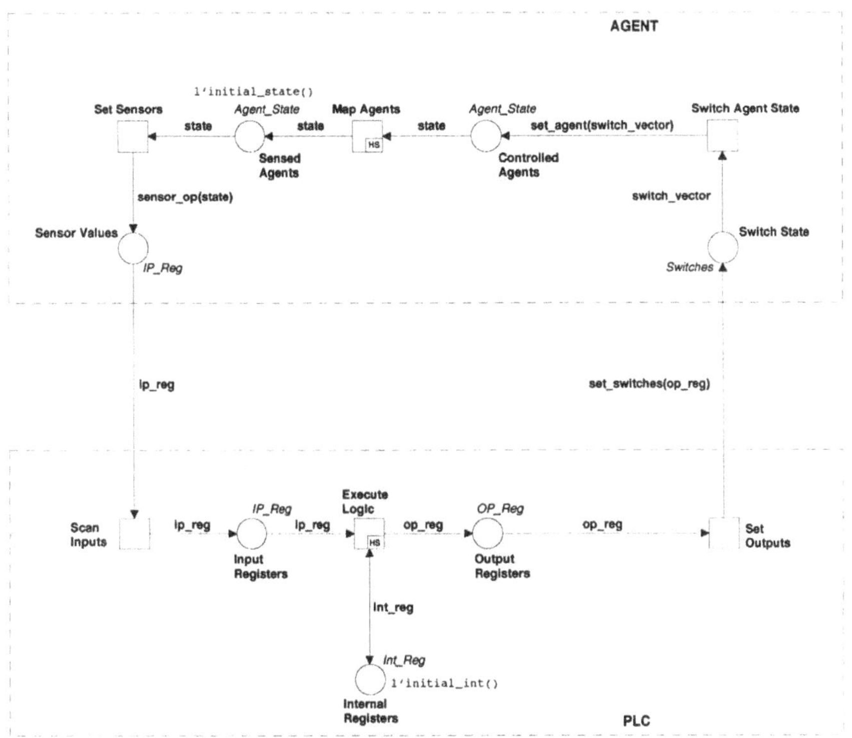


**Figure 9** The Control Loop CP-net

bottom-up modelling made use of existing system control logic. The fusion process, summarised in Section 4.4, yields a system model which contains a hierarchy of behavioural requirements *and* a set of control logic specifications, so making it easier to identify the relationships between the bottom-level control logic specifications and the higher level system behavioural requirements. The fused CP-net model is therefore a useful starting point from which to derive an error-free assembly line control specification.

# REFERENCES

[Billington, 1991] Billington, J. (1991). FORSEEing Quality Telecommunications Software. *1st Aust. Conf. on Telecommunications Software* pp. 169–74.

[Darabi and Jafari, 1994] Darabi, H. and M.A. Jafari (1994). A Zero-One Programming of Petri Nets to Coloured Petri Net Transformation. *Proc. 4th Int. Conf. on CIM and Automation Technology* pp. 25–31.

[Desrochers and Al-Jaar, 1995] Desrochers, A.A. and R.Y. Al-Jaar (1995). *Applications*

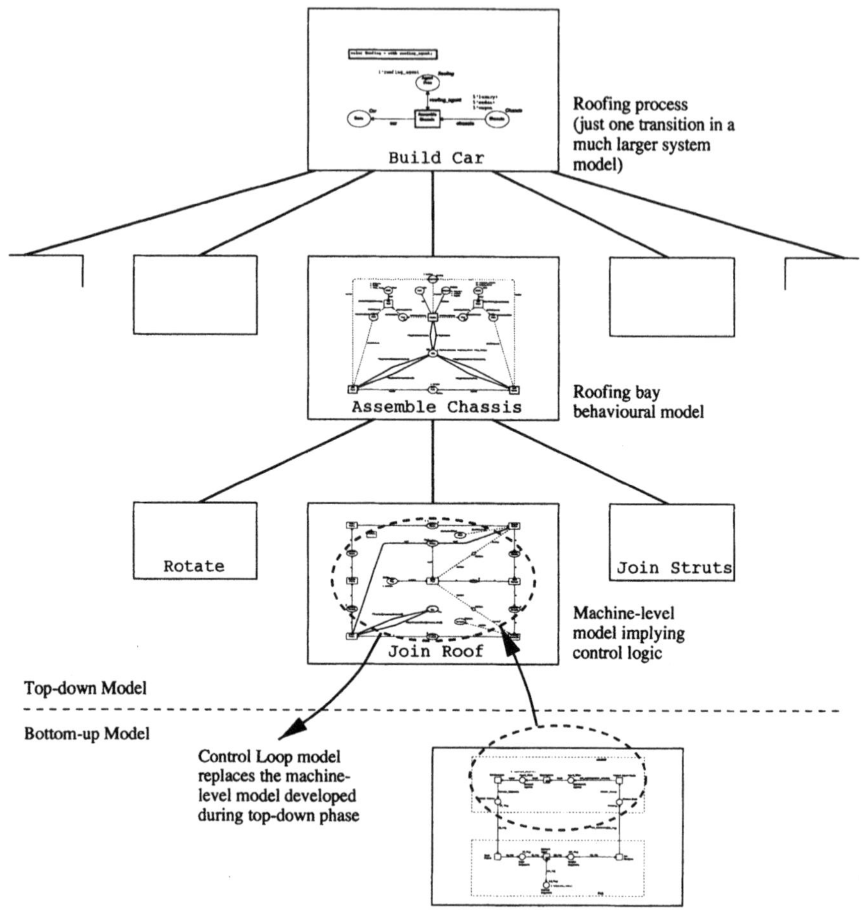


**Figure 10** Fusing the Top-Down and Bottom-Up models

*of Petri nets in manufacturing systems: modeling, control, and performance analysis.* IEEE Press.

[Farrington and Billington, 1996] Farrington, M. and J. Billington (1996). Modelling an Automotive Assembly Line. *Proc. 13th IFAC World Congress.*

[Gilles, 1990] Gilles, M. (1990). *Programmable Logic Controllers: Architecture and Applications.* John Wiley & Sons. pp. 1–20, 165–213.

[Jensen, 1992] Jensen, K. (1992). *Coloured Petri Nets Basic Concepts, Analysis Methods and Practical Use.* Springer-Verlag.

[Lester, 1994] Lester, L.N. (1994). *TORAS Documentation. Final documentation for the TORAS development contract.* UNICO Computer Systems.

[Reisig, 1992] Reisig, W. (1992). *A Primer in Petri Net Design.* Springer-Verlag.

[Venkatesh *et al.*, 1994] Venkatesh, K., M.C. Zhou and R. Caudill (1994). Automatic Generation of Petri Net Models from Logic Control Specifications. *Proc. 4th Int. Conf. on CIM and Automation Technology* pp. 242–247.

[Venkatesh, *et al.*, 1994] Venkatesh, K., M.C. Zhou and R. Caudill (1994). Comparing Ladder Logic Diagrams and Petri Nets for Sequence Controller Design Through a Discrete Manufacturing System. *IEEE Trans. on Industrial Electronics* vol. 41, no. 6, pp. 611–619.

[Zhou and Leu, 1991] Zhou, M.C. and M.C. Leu (1991). Petri net modeling of a flexible assembly station for printed circuit boards. *Proc. IEEE Int. Conf. on Robotics and Automation* pp. 2530–2535.

[Zhou *et al.*, 1989] Zhou, M.C., F. DiCesare and A.A. Desrochers (1989). A top-down approach to systematic synthesis of petri net models for manufacturing systems. *Proc. IEEE Int. Conf. on Robotics and Automation* pp. 534–539.

6

# Modeling and Simulation of Combined Discrete Event-Continuous Systems Using DEVS Formalism and Object-Oriented Paradigm

*M.TEGGAR and R. SOENEN*
*G.I.L -LAMIH UVHC*
*BP 311 , Le Mont Houy ,59326 - Valenciennes Cedex*
*Tél.: 27.14.13.47, E-mail: teggar@univ-valenciennes.fr*

**Abstract**

This paper presents an approach to a simulation-based design methodology and is motivated by the need to provide models representation schemes and simulation software tools for combined discrete event-continuous systems. Our approach is based on an extension of the *DEVS (Discrete EVent System Specification)* introduced by Zeigler, to include a combined discrete event-continuous systems. We show how the interactions between the discrete-event part and the continuous part and their consequence on the system behavior can be expressed. By employing the Object-Oriented paradigm, the extended model is implemented as an abstract class, in which the methods describe the model's behavior and data represent its variables.

**Keywords**

**Modeling and simulation, combined simulation, discrete event simulation, continuous system simulation**

## 1 Introduction

In addition to discrete event models and continuous models, combined discrete-event-continuous modeling and simulation is finding ever more application in the analysis and design of complex manufacturing systems. A lot of complex systems, such as steel-making, metallurgy, chemical industry ..., involve both processes which are easily described by differential equations and processes which are better described by models whose changes in state occur at discrete instants rather than continuously in time.

Classical approaches to simulation typically use:

- differential equations based models and frequency models for continuous systems (continuous states, continuous time),
- state-transition, automaton and Markovian models for discrete-time and discrete-events dynamic systems.

Unfortunately, these different models are incompatible because they deal with variables which don't belong to the same mathematical space, and the time variable is not used in an identical manner.
In fact, the same physical system may be modeled in an entirely continuous or an entirely discrete fashion depending on the modeler point of view and the duration for which the system will be observed. However, many works, (Wang, 89) ,(Cellier, 79), (Fahrland, 70), show that there exist classes of problems which can not be modeled satisfactorily by either a purely discrete or a purely continuous formulation. And the choice between the two points of view is not always easy, since any system optimization study must be global and the continous part and the discrete part of a process are often closely linked (Caristi, 91).

Although many formalisms have been developed to study, on the one hand discrete event dynamic systems and on the other hand, continuous system, there is no formulation that meets them all. Indeed, each of these various formalisms can be viewed as offering a potential application benefit based on the particular form of abstraction implicit in its "*world of view*". However, since reality does not usually constrain itself to one such "*world view*", the need to formulate representation schemes which allow to bridge the gap between the two description form becomes essential.

Two main approaches to the modeling of combined systems have recently emerged :
- defining a single formalism which encompasses the discrete event and continuous behavior and which uses a homogeneous model structure (Praehofer, 90, 92) (Zeigler, 89), (Buisson, 93)
- using a specific formalisms for each class of system components and defining an appropriate interface between them (Stiver, 93) (Antsaklis, 93) (Alla, 94).

The DEVS formalism, initially developed for formalizing simulation models of discrete event systems to be simulated, supports an open approach for the exploration of new simulation-based representations for many classes of systems. It provides a mechanism for the specification of simulation models in a modular and hierarchical manner and represents an alternative to the traditional worldviews of discrete-event-simulation languages.

To shed some light on the *DEVS* formalism (Zeigler, 76), a conceptual framework underlying the formalism is first described. We illustrate how basic models are specified and how these models are connected together in hierarchical fashion to form complex models. Then we show how continuous states are introduced to extend the *DEVS* expressibility to combined discrete-continuous systems. Thus, the extended scheme represents both types of system and the interactions between them in the same modeling framework.

After developing a mapping of the extended scheme onto a generic description so that classes of objects form a *taxonomical hierarchy* in which they are arranged according to their degree of generality, we describe an implementation of the simulation strategy using the *abstract simulators* principles (Concepcion, 88) applied to discrete event-continuous simulation.

## 2 THE DEVS FORMALISM

*DEVS (Discrete EVent System Specification)* (Zeigler, 76, 84, 87) is a set-theoretic based formalism that provides a system theoretically grounded means of expressing hierarchical and modular discrete event simulation. It is the shorthand formulation needed to specify systems whose input, state and output trajectories are piecewise constant.

### 2.1 *Basic model*

In this formalism a basic model (called atomic model) is defined by the structure:

$$M = < X, S, Y, \delta_{int}, \delta_{ext}, \lambda, ta >,$$

where
$X$ = set of external input event
$S$ = set of sequential states
$Y$ = set of output
$\delta_{int} : S \rightarrow S$ is the internal transition function
$\delta_{ext} : Q \times X \rightarrow S$ is the external transition function
where $Q = \{ (s,e) \mid s \in S, 0 \leq e \leq ta(s) \}$ is the total state set
$\lambda : S \rightarrow Y$ is the output function
$ta : S \rightarrow R_0^+$ the time advance function.

There are two kinds of events; external events (input events) and internal events which are time scheduled events. For each state $s$ the time advance function $ta$, which is a mapping from state space $S$ to positive real numbers, defines the time interval to the next internal event. When this time given by $ta(s)$ has elapsed, an internal event occurs. The system produces an output $\lambda(s)$ and the internal transition function specifies the next state $s' = \delta_{int}(s)$ to which the system will transit. If an external event (an input) occurs at elapsed time $e$ which is less than $ta(s)$ time units, then a new state $s'$ is computed by means of the external transition function $\delta_{ext}(s, e,x)$.

### 2.2 *Multicomponent Model*

Several atomic models can be coupled to form a multicomponent model defined by the structure:

$$DN = < D, \{M_\alpha\}, \{I_\alpha\}; \{Z_{\alpha,\beta}\}, select>$$

where
$D$ = set of component names,
For each $\alpha$ in $D$
$M_\alpha$ = the component model
$I_\alpha \subseteq D$ set of influences of $\alpha$
and for each $\beta$ in $I_\alpha$
$Z_{\alpha,\beta}$ is the $\alpha$-to-$\beta$ output translation function

*select* : $2^D \rightarrow D$ is the tiebreaking selector function. It selects a component from the imminent components having the minimum next event time. This component is then allowed to execute its next event transition.

A coupled model *DN* can be expressed as an equivalent atomic model and can itself be employed in a larger coupled model. This shows that the *DEVS* formalism is closed under coupling and hence the construction of modular hierarchical models is possible and formally well defined.

# 3 EXTENDED DEVS EXPRESSIBILITY TO CONTINUOUS STATES SYSTEMS

In this section, we introduce an extension of *DEVS* formalism to combined systems whose input and output are eventlike. The systems contain both discrete states and continuous states and the dynamic behaviour of the continuous components can be described by differential equations.

## *3.1 Example : The Bottle-filling process*

Consider a cylindrical tank with two threshold sensors at two levels called *L* (Low) and *H* (High) and a bottle conveyer as it shown in the figure 1 below ;

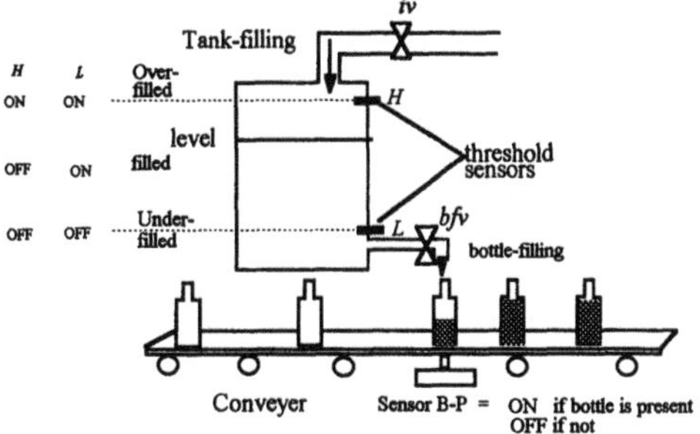

Figure 1. Bottle -filling example.

The filling process receives three types of event:

- $e_{11}$ / $e_{12}$ : Open_influx_valve/ Close_influx_valve ;
- $e_{21}$ /$e_{22}$ : Stop_on_error/ Re-Start; (from the higher level Control System.)
- $e_3$ : Bottle_Arrived ; (from the conveyer when an empty bottle arrive under the tank's outlet.)

And it transmits

- an event $o_1$=*move_conveyer* when a bottle is full ( to the conveyer)
- and two events $o_2$ =*Underfilled* and $o_3$ =*Overfilled* (to the higher level control system.)

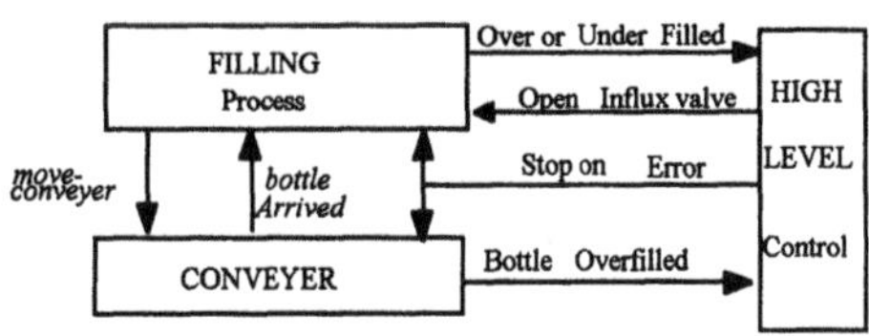


Figure 2. A coupled discrete event-continuous system.

When the filling process receives an event $e_{11}$ the influx valve (*iv*) is fully open and the tank is filled with constant flow $F_{in}$. *iv* is closed when $e_{12}$ is received.When it receives an event $e_{22}$; this means that there is an error (bottle overflow, a failure of sensors, a leak in the tank,...) then the filling process is stopped. Event $e_3$ indicates a bottle-arrival, then the bottle-filling-valve (*bfv*) is fully open; the flow out is given by $F_{out}(t) = R.\ l(t)$ (*R* is the fill rate, some positive real number and $l\ (t)$ is the level of the liquid in the tank). When the volume of the bottle is reached the *bfv* is closed and an event *move_conveyer* is sent to the conveyer.

## 3.2 Extended DEVS Model

The example shows that, in addition to the sequential states, the total state contains some continuous variables (Figure 3) (the level of the liquid in the tank and the volume out in the filling process).

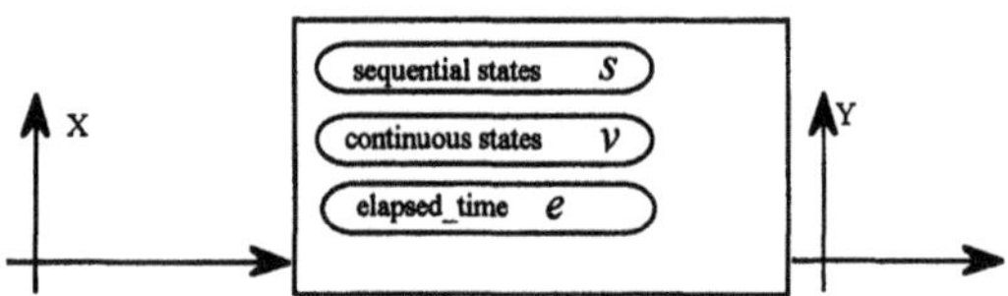


Figure 3 - Combined Discrete event-continuous system.

Moreover, two types of internal event are distinguished :

- The state-events, which are not time scheduled events but are caused by the continuous changes of the continuous states.
- The time events, which are time scheduled as in the DEVS definition given above.

A Combined Dynamic System is a tuple $< X, Q, Y, E, \delta, \mathcal{F}, \lambda, \sigma, ta >$
where :

- $X$ is a finite set of input events and $Y$ is a finite set of output events
- $E$ is a finite set of internal state-events;
- $Q$ is the state set range defined by : $Q = S \times R^n \times R^+$ , $S$ is discrete state set and $R^n$ is a continuous state-space;

$$Q = \{q = (s, v, e) \mid s \in S,\ v \in R^n,\ e \in R^+_\infty\}$$

- The conditions of occurrence of internal state-events are given by the function $\sigma$ which associates at each $e_s \in E$ a predicate $P(q)$ where $q$ is the total state variable and $P(q)$ is a logical expression

  Whenever the condition $P(q)$ <u>becomes</u> true, an internal state-event occurs.
- An internal time event $\varepsilon$ occurs if the condition $e \geq ta\ (s)$ becomes true
- $\delta$ is the discrete state transition defined by the two functions $\delta_{int}$ and $\delta_{ext}$ as follows :

$$\delta_{ext} : Q \times X \rightarrow Q;$$
$$\delta_{int} : S \times E \cup \{\varepsilon\} \rightarrow Q;$$

- $\mathcal{F}$ is the differential equation defining the rate of change of the continuous state variables;
- $\lambda : S \rightarrow Y$ is the output function

### 3.3 The Filling-Process Model

For the filling process described above, the sets $X$, $Y$ are defined by :

$X = \{e_{11}, e_{12}, e_{21}, e_{22}, e_3\}$;

$Y = \{o_1, o_2, o_3\}$;

*Let* $s = \begin{bmatrix} L \\ H \\ bfv \\ iv \\ phase \end{bmatrix}$ denote the sequential state and $v = \begin{bmatrix} l \\ v\_out \end{bmatrix}$ denote the continuous state

where : $L, H, phase \in \{ON, OFF\}$; $phase = OFF$ if the system is stopped ; $ON$ otherwise;

$bfv$, $iv \in \{0, 1\}$, the two variables take the value 0 when the corresponding valve is closed and 1 if it is fully open

the total state set is

$$Q = \{q = (s^T, v^T, e)\} = S \times R^2 \times R^+_\infty \; ; \; S = \{ON, OFF\} \times \{ON, OFF\} \times \{0,1\} \times \{0,1\} \times \{ON, OFF\};$$

The set $E$ of the state-events contains five events : $E = \{ e_{s_1}, e_{s_2}, e_{s_3}, e_{s_4}, e_{s_5} \}$

and $\sigma$ is given by :

$$\begin{aligned}
e_{s_1} &\mapsto l \geq HIGH \\
e_{s_2} &\mapsto l < HIGH \\
e_{s_3} &\mapsto l > LOW \\
e_{s_4} &\mapsto l \leq LOW \\
e_{s_5} &\mapsto v_out \geq B_volume
\end{aligned}$$

$ta(s) = \infty$, $\forall\ s \in S$ *No time event*

The external transition function $\delta_{ext}$ is :

$$\begin{aligned}
&((L, H, bfv, 0, ON), v, e, e_{11}) \mapsto ((L, H, bfv, 1, ON), v, 0) \\
&((L, H, bfv, 1, ON), v, e, e_{12}) \mapsto ((L, H, bfv, 0, ON), v, 0) \\
&((L, H, bfv, iv, ON), v, e, e_{21}) \mapsto ((L, H, 0, 0, OFF), v, 0) \\
&((L, H, bfv, iv, OFF), v, e, e_{22}) \mapsto ((L, H, 0, 0, ON), v, 0) \\
&((L, H, 0, iv, ON), v, e, e_3) \mapsto ((L, H, 1, iv, ON), v, 0)
\end{aligned}$$

the internal transition function is defined by :

$$\begin{aligned}
&((L, OFF, bfv, iv, ON), e_{s_1}) \mapsto ((L, ON, bfv, iv, ON), v, 0) \\
&((L, ON, bfv, iv, ON), e_{s_2}) \mapsto ((L, OFF, bfv, iv, ON), v, 0) \\
&((OFF, H, bfv, iv, ON), e_{s_3}) \mapsto ((ON, H, bfv, iv, ON), v, 0) \\
&((ON, H, bfv, iv, OFF), e_{s_4}) \mapsto ((OFF, H, bfv, iv, ON), v, 0) \\
&((L, H, 1, iv, ON), e_{s_5}) \mapsto ((L, H, 0, iv, ON), (1, 0), 0)
\end{aligned}$$

The differential equations are given by :

$$\mathcal{F} = \begin{cases} \dfrac{dl}{dt} = \dfrac{1}{TS} \cdot \left[F_{in}.i\_v - R.l(t).b\_f\_v\right]; & \text{where } TS \text{ is the cross-sectional area of the tank} \\ \dfrac{dv\_out}{dt} = R.l(t).b\_f\_v & \end{cases}$$

the output function :

$\lambda$ (*OFF* , *OFF* , *bfv*, *iv*, *ON*) = ( $o_2$ , H_L_C); *Tank_underfilled* (to Higher level control)

$\lambda$ (*ON* , *ON* , *bfv*, *iv*, *ON*) = ( $o_3$, H_L_C) ; *Tank_overfilled* (to Higher level control)

$\lambda$ (L , H, 0, *iv*, *ON*) = ( $o_1$ , Conveyer) ; *move_conveyer* (to CONVEYER)

$\lambda$ ( s) = $\phi$ ; for the other cases ($\phi$ = no event)

# 4 SIMULATION WORLD VIEW

The *Object-Oriented Concept* is not only a useful programming style and an efficient way to organize software systems, but it is also regarded as a powerful paradigm that can serve as a useful way to express computational models for large, complexly interacting systems. Encapsulation, inheritance, abstraction and evolution of individual objects support concurrency, incremental modifiability and reusability of models. Thus, it provides a common basis that can be used to implement modular simulation programs.

The simulation strategy for DEVS models is based on the abstract simulator principles developed by Zeigler et al (Zeigler, 84) (Concepcion, 85, 88) (Kim, 88, 89). The abstract simulator includes the algorithmic interpretation of the dynamic behaviour implicitly specified by the model.

The abstract simulator has the same structure as the hierarchical structure of the DEVS models. Thus the model can be directly transformed into an executable simulation program using the object oriented concept. A full explication and the class specialization hierarchy are given in (Zeigler, 90).

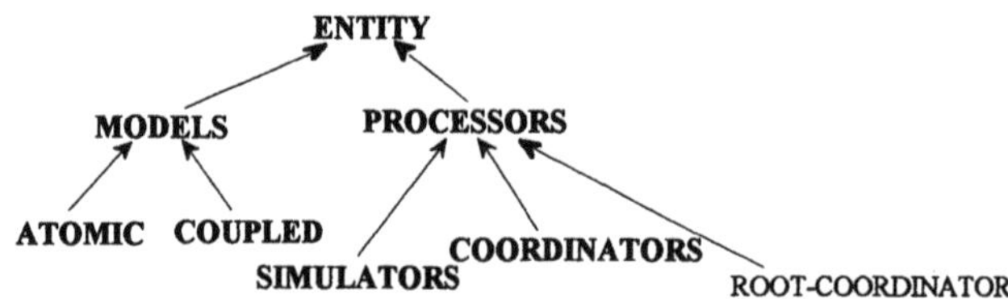


Figure 4. The abstract classes hierarchy.

The root class *ENTITY* provides general definitions for data and methods of general utility. The class *MODELS* which in turn is specialized into more specific classes *ATOMIC* and *COUPLED* provide the basic constructs needed to specify a model's variables (Instance variables) and its behavior (methods). The class *PROCESSORS* and its subclasses implement methods to carry out the simulation and compute the states and outputs trajectories.

Using the same methodology, we have developed an implementation using the C++ language according to the class hierarchy shown in figure 4 (Teggar, 96). Since these classes are defined as abstract classes, they are implemented in generic form as template classes. The methods expressing a specific behavior of user model are defined as virtual functions with template data types.

## *4.1 The class combined model*

The inheritance mechanism supports incremental specification and composition of behaviour so that user methods are defined as subclasses, which inherit the code representing data and behaviour common to the other models. The class for *Combined models* is a specialization of the class *ATOMIC* in which we can add new instance variables for the continuous states and new methods :

- *sigma* : which returns a state event identifier when the associated condition becomes true and 0 otherwise
- *derivative* : computes the derivative values of the continuous states .

Using pseudo-code, we describe the class of the filling process model as a subclass of *combined_models.*

```
class Filling_Process : Combined_model
 ClassVars
  LOW , HIGH    :real
  B_volume      :real
 InstVars
   L  , H ,phase         :[ON , OFF]
   bfv , iv              : 0..1
   l ,Dl , v_out ,Dv_out : real
   e                     : real

 Methods
```

```
Int_Trans ( event es)
  case es of
      es1 : if (H=OFF) and
                (phase = ON)then
              H := ON;
              e := 0
            else error
      es2 : if (H=ON) and
                 (phase = ON)then
              H := OFF;
              e := 0
            else error
      es3 : if (L=OFF) and
                 (phase = ON)then
              L := ON;
              e := 0
            else error
      es4 : if (L=OFF) and
                 (phase = ON)then
              L := ON;
              e := 0
            else error
      es1 : if (bfv=1) and
                 (phase = ON)then
              bfv := 0;
              e := 0
              v_out := 0
            else error
end
```

```
Ext_trans ( event x)
   case x of
       e11 : if (phase = ON) and
                   (iv = 0) then
                 iv := 1
                 e := 0
       e12 : if (phase = ON) and
                  (iv = 1) then
                 iv := 0
                 e := 0
       e21 : if (phase = ON) then
                 iv := 0
                 bfv := 0
                 phase := OFF
                 e := 0
       e11 : if (phase = OFF) then
                  phase := ON
                  e := 0
       e3 : if (phase = ON) and
                (bfv = 0) then
                 bfv := 1
                 e := 0
end
```

```
Derivative ( )
   Dl :=((Fin*iv )-(R* l * bfv ))/TS
   Dv_out := bfv * R * l
end
```

```
Sigma ( )
      if (l>= HIGH ) then
        return (es1)
      if (l< HIGH ) then
        return (es2)
      if (l> LOW ) then
        return (es3)
      if (l =< LOW ) then
        return (es4)
      if ( v_out >= B_volume) then
        return (es5)
return (0)

end
```

```
Output_func ( )

      if (L = OFF) and (H = OFF)
         and (phase = ON) then
        send ( O2 , H_L_C)
      if (L = ON)and (H = ON)
        and (phase = ON) then
        send (O3 ; H_L_C)
      if ( bfv = 0)
        and (phase = ON) then
        send (O1 , CONVEYER)

end
```

The specialization gives the specific definitions by overriding the virtual methods *Int_Trans, Ext_Trans* and *Output_Func,* of the class atomic and *Derivative* and *Sigma* of the class *Combined_model*

## 4.2 Abstract simulator principles

The abstract simulators are defined as the interpretation of dynamic behaviours specified by models. Essentially, they are objects whose methods are an algorithmic description of how to carry out the instructions implicit in models to generate their behavior. Simulators are assigned to handle basic models in a one-to-one manner. The whole simulation program is then represented by a network of such *(model, simulator)* pairs.

There are two types of abstract simulator: Simulators and Coordinators; assigned to handle atomic-DEVS models, and coupled models in a one-to-one manner respectively. Simulation proceeds by messages that carry data and synchronization information passed among the abstract simulators.

### 4.2.1 The. Abstract Simulator of Discrete Event System

The operation of an abstract simulator involves four types of messages : * , x, y and done-messages. It consists of two methods that handle four main variables :

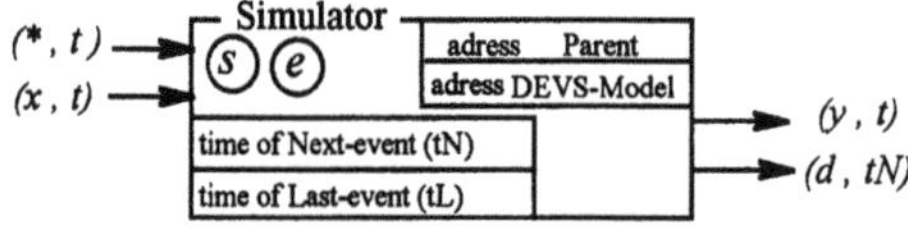


Figure 5. Discrete Event Simulator

- s : is the sequential state variable,
- e : the elapsed time since the last state transition,

-tL : the time of the last event and
-tN : is the time of the next internal state transition.
The algorithm of the abstract simulator is divided onto two parts : One method is activated on a reception of an (x,t) message, and the other one corresponds to an (*,t) message.

```
When receive an input (x,t)
      if tL <= t <= tN then
        e := t-tL
        s := δext (s e x)
        tL := t
        tN := tL + ta (s)
        send (d,tN) to co-ordinator
        else error
end

When receive an input (* ,t)
      if t = tN then
        y :=  λ (s)
        send (y,t) to co-ordinator
        s := δint (s)
        tL := t
        tN := tL + ta (s)
        send (d,tN) to co-ordinator
        else error
end
```

In (Zeigler, 1984), the abstract simulator was shown to satisfy the criterion of correct simulation of a DEVS model. The *-message indicates that an internal event shall be executed and is transmitted to the imminent component. If there are more than one imminent component, the tie breaking function Select is used to select one of them. When a *-message is received by a simulator, it sends its output as a y-message and carries out the internal transition in the associated DEVS model. The output (y-message) is sent back to the parent coordinator which consults the external output coupling and the internal couplings to obtain the addresses to which the message should be sent as an x-message. When a Coordinator receives an x-message, it consults the external input coupling to generate the appropriate x-message for the subordinate influenced by the external event. When a Simulator receives an x-message, it executes the external transition of the associated DEVS. A done-message indicates the completion of the state transition and contains the time of the next internal transition.

### 4.2.2 The Abstract Simulator for Combined Model

The simulators described in the preceding paragraph, are defined to handle discrete event systems. So to simulate a combined discrete event-continuous state system, we must define a special simulator which has methods to compute the continuous state values during the observation time.
The extension retains the basic structure of the abstract simlators. Since the state-events are not time scheduled events, a simulator for a combined model can not determine the time of the next internal transition. To solve this problem, we use a sampling method which predicts the occurrence of an internal event with an estimated time *tNe*. If a state event occurs at time $t < tNe$, a time warp mechanism is used to correctly simulate a multicomponent model. The mechanism to rollback an object is the heart of the time warp. The time warp exerts no effort

to ensure that messages are delivered to an object in increasing timestamp order. Instead, the simulation proceeds on the assumption that there will be no stragglers (messages with time stamps less than the local time ). But if a straggler should arrive at time ***t***, the object must roll back to the time ***t*** and cancel all the side effects that occurred as a result of processing messages with timestamps greater than ***t*** (Jefferson, 85 ,87, 89).

```
When receive an input (x,t)
      if tL <= t <= tN then
        compute the continuous state and update e from tL
        until t or tE the time of state-event
        if e = t -tL then
        s := δext (s e x)
        tL := t
        tN := tL + ta (s)
        send (d,tN) to co-ordinator
        else
            send (anti-message, tN)to co-ordinator
            send (d,tE)to co-ordinator
        else Rollback (Time Warp mechanism)
end

When receive an input (* ,t)
      if t = tL then
        y :=  λ (s)
        send (y,t) to co-ordinator
        s := δint (s)
        tL := t
        tN := tL + ta (s)
        send (d,tN) to co-ordinator
        else if t = tN
        compute the continuous state and update e from tL
        until t or tE the time of state-event
        if there is a state-event (tE)
             y :=  λ (s)
             send (y,t) to co-ordinator
             s := δint (s)
             tL := tE
             tN := tL + ta (s)
      else Rollback
end
```

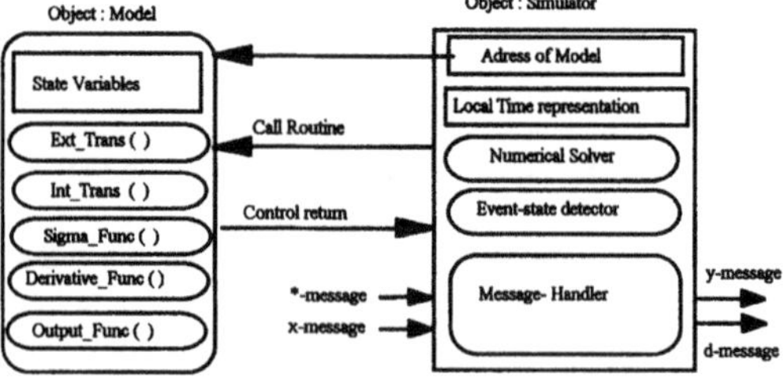


Figure 6. A model -Simulator pair.

Note that:

- The messages exchanged among the abstract simulators are timestamped.
- When a simulator of combined model receives an *(x, t)* or *(* , tNe)* message, before executing the external or internal transition it updates its local clock by employing the solver based on a numerical integration method to compute the continuous states value step by step and the event detector method to check the value of the sigma function. If a state event occurs at time $tN < tNe$ , the simulator transmits an *(* , tN)* to the parent coordinator (figure 7).
- When a Coordinator receives a message with a timestamp *tN* less than its local clock, it releases the rollback mechanism by means of the *antimessages*, and start again with the new value *tN* of the internal state-event.

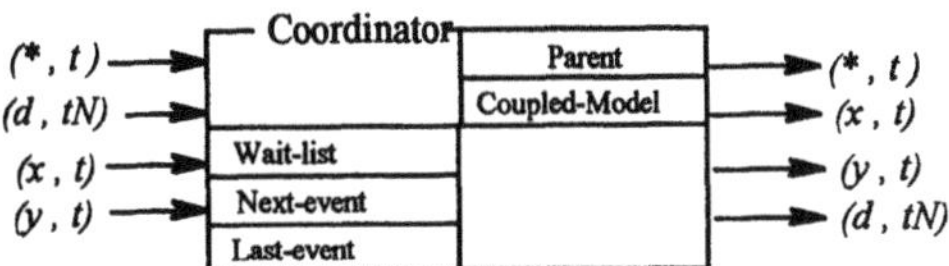


Figure 7. The Coordinator of Coupled Model.

## 4 CONCLUSION

A variety of paradigms are being actively explored to deal with systems where a large number of discrete events and continuous dynamic behaviours are simultaneously important. An extension of the DEVS formalism shows how it is possible to express models of combined discrete event-continuous systems, so that the whole formalism guarantees the strength of the theoretical based methods for simulation. By the bottle-filling process example, we have illustrated that the proposed modeling framework makes a clear description of complex combined systems possible. The object oriented paradigm provides convenient tools to carry out this abstract formalism concept and improves the isomorphic relation between the model and its representation in the simulation program.

## REFERENCES

Alla H. (1994) Les Réseaux de Petri: des Evolutions Particulièrement Adaptées aux Systèmes Hybrides. in ADPM94, Les Systèmes Dynamiques Hybrides.

Buisson J. (1993) Analysis and Characterisation of Hybrid Systems with Bond-Graphs. in IEEE International Conf. on Systems, Man and Cybernetics, Vol 1.

Caristi J. and Sands D.C. (1991) Simulation of Epiphytic Bacterial Growth under Field Conditions Simulation J..

Cellier F.E. (1979) Combined Continuous/Discrete System Simulation : Languages, Usefulness, Experiences and Future Development in Methodology in Systems Modelling and Simulation, Eds. B.P. Zeigler, M.S. Elzas, G.J. Klir & T.I. Ören, North Holland Pub.,.

Concepcion A.I.(1985) The Implementation of the Hierarchical Abstract Simulator on The HEP Computer. Proc. Winter Simulation Conf.

Concepcion A.I. and Zeigler B. P.(1988) DEVS Formalism: A Framwork for Hierarchical Model Devlopement: IEEE Transaction on Software Engineering, Vol. 14, NO. 2.

Fahrland D. (1970) Combined Discrete Event Continuous Systems Simulation. Simulation J..

Jefferson D. R. and Sowizral H. (1985a) Fast Concurrent Simulation using the Time Warp Mechanism. In Proc. SCS conf. on Distributed Simulation, San Diego.

Jefferson D. R. (1985b) Virtuel Time. ACM Transaction on Programming Languages and Systems, Vol. 7, No. 3.

Jefferson D. R., Backman B.and al (1987) The Time Warp Operating System. 11th Symp. Operating Systems Principles, Vol 21.

Jefferson D. R. and al (1989) The Performance of a Distributed Combat Simulation with the Time Warp Operating System. Concurrency Pratice and Experience, Vol. 1, No. 1.

Kim T.G. (1988) Knowledge-based environment for hierarchical modelling and simulation. Ph. D. dissertation, University of Arizona, Tucson, AZ.

Kim T.G.and Zeigler B.P. (1989) ESP-Scheme : A Realization of System Entity Structure in a LISP Environment. Proc. AI and Simulation Multiconference, SCS Publications, San Diego.

Praehofer H. (1990) System Theoretic Formalisms for Combined Discrete-Continuous System Simulation. Int. J. General Systems, Vol 19, 219-240.

Praehofer H. (1992) An Environment for DEVS-Based Multiformalism Simulation in Common Lisp. Special issue on : Software Environment for Discrete-Event Dynamic Systems.

Stiver J.A. and Antsaklis P.J. (1993) On the Controllability of Hybrid Control Systems. in 32nd IEEE Conf. on Decision and Control.

Teggar M. (1996) Modélisation et Simulation des Systèmes Dynamiques Hybride. Thèse de Doctorat de l'université de Valenciennes et du Hainaut Combresis.

Wang Q. (1989) Management of Continuous Models in DEVS-Scheme: Time Windows for Event-Based Control Masters Thesis, Dept. of Electrical an Computer Engineering, University of Arizona, Tucson, AZ.

Zeigler B.P. (1976) Theory of Modelling and Simulation. J. Wiley NY . (Reissued by Krieger Pub. Co., Malabar, FL, 1985) .

Zeigler B.P. (1984) Multifacetted Modelling and Discrete Event Simulation. Academic Press, London.

Zeigler B.P. (1987) Hierarchical, Modular Discrete Event Simulation in an Object-Oriented Environement. Simulation Journal, Vol. 49:5.

Zeigler B.P. (1989) DEVS Representation of Dynamical systems : Event-Based Intelligent Control. Proceedings Of The IEEE, Vol. 77, NO. 1.

Zeigler B.P. (1990) Object-Oriented Simulation with Hierarchical, Modular Models. Academic Press, London.

# 7

# Principles of CASE Tool design for Automation Control

*Dr. Wolfgang Brendel*
*infoteam Software GmbH*
*D-91088 Bubenreuth, Am Bauhof 4, Germany*
*phone +49-9131-78 000, fax +49-9131-78 0050*
*e-mail 100024.1231@compuserve.com*

**Abstract**

Twelve principles which should be obeyed by every designer of CASE-tools for control automation are postulated, in order to avoid that a post-modern computer scientist with fancy ideas neglects the requirements of users who are much more conservative regarding taste, attitude and practicability.

Much more importantly, PLC-programming has to be different from conventional software development because of the different kind of people who are involved in that task. The purpose of the following statements is to give some hints as to which principles should be obeyed in designing a PLC-Programming system.

Giving examples from the implementation of the Open Development Kit, an interactive PLC-programming support environment conform to IEC 1131-3 , we also give guidelines as to how to implement these principles in a real world environment. We close with first experiences from several hundred users and give an outlook to further topics both in research and development for the near future.

**Keywords**

IEC 1131-3, PLC, programming, CASE-tools, OPEN DK

## 1 INTRODUCTION

To enhance the performance of the Software Engineering Process, it is required to shorten the development cycle of new automation control software. This aim is achieved by introducing engineering processes which apply different engineering methods in several phases of the overall design and the development cycle. This is accompanied with the development of CASE-Tools supporting these methods. Nowadays we have a huge bundle of design tools, graphical editors, interactive workbenches, incremental compilers and on-line debugging utilities with one major obstacle: the lack of common interfaces, ensuring a smooth transition from design down to operation and ensuring reliability of an automated manufacturing system.

That's why we now have to pursue two objectives:

- to reduce the number of tools involved as far as possible and
- to standardize the interfaces between these tools.

The first target can be reached by incorporating tasks like code generation in design tools used at an early stage of the engineering process, eliminating the need for the user to occupy himself with compilers and generated code in detail.

To reach the second aim - to standardize interfaces between heterogeneous tools - is much more difficult, because it requires the co-operation of tool suppliers who often are competitors. Nevertheless we know from all sectors of engineering that international standards are mandatory for technical progress. So there is no alternative but the standardization of interfaces.

## 2 CASE-TOOL ARCHITECTURE FOR CONTROL AUTOMATION

Which CASE-Tool design will meet these objectives? The ultimate judge will be the engineer who has to perform a most complicated task with a fixed deadline: the day the manufacturing system he designs has to be up and running!

In order to design a flexible architecture of software tools for control technology, infoteam made an analysis of user requirements in the field of PLC-programming and control automation. Based on these results, we found out that principles can be stated for the development of CASE-Tools which shall ensure that a software tool will match the requirements of the end-user and, at the same time, will fulfill the two overall objectives as mentioned above.

The key to an open architecture for control automation tools is to define a common framework in which heterogeneous tools can cooperate and to develop an Application Programmer Interface - the so called Open-Software-Link „OSL“ used by developers of tools for automation control programming.

The Software-Industry market is dominated by de facto industry standards in areas like operating systems and database technology as well as programming languages and compiler design. This eases exchange of information and technology.

Industrial Control Automation is one of the few exceptions: Until recently there was no world-wide standard at all. Each PLC-manufacturer had its own style of programming tools and even worse, depending on the country of the company each control system was programmed in a different programming language:

- in France, Spain and Italy using GRAFCET
- in the United Kingdom and the United States using Ladder Diagrams
- in Germany, Austria and Scandinavia using Function Block (FB) Diagrams
- and in all countries some dialects of Instruction List and other languages.

> *Principle 1:*
> *Don't try to convince the user of your favourite programming methodology or programming language. Support the* ***method or language*** *the user wants to use.*

Fortunately the emerging international standard IEC 1131-3 defines all five languages in a consistent way. Moreover, the user and vendor organization PLCopen defines compliance rules that have to be fulfilled by the corresponding tools implementing that standard. This is a prerequisite for the standardization of widely accepted interfaces between CASE-tools, interactive editors, compilers, MMI-tools and a common database definition all used throughout the life-cycle of an Industrial Control Automation Project.

The lack of standardized interfaces and common data definitions led to the circumstance that the wheel is reinvented more than once in such an automation project. But even worse, mistakes are introduced and lead to poor quality and schedules which are overdue in most automation projects. Because there is no way to cooperate between tools from different vendors each company invented their own programming support environment.

Users are therefore adamant about the use of standard PCs for programming and hail the standardization of the programming languages by IEC 1131-3 as the breakthrough they have been waiting for.

The logical consequence is the demand for new manufacturer-independent programming tools in accordance with IEC 1131-3, which can be used on any PC to produce machine code for (almost) every PLC and put it into operation.

Working on this assumption and years of experience, the Open DK portable programming system was developed following the principles:

- Using standardized components, such as editors, compilers and management tools
- Adopting components where different requirements ask for different solutions
- Filling in missing pieces by developing special tools or enhancing already existing ones.

There are clearly defined interfaces between all parts of the software, so that they can easily be exchanged for new versions because the interface specifications are upward compatible. In this way, new applications can be added to the system at any time.

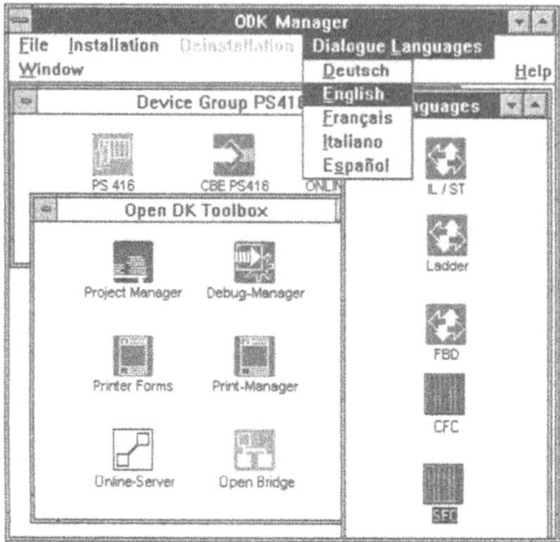


Figure 1: Architecture of Open Development Kit

The majority of PLC systems today, however, provide little or no support for the features of the IEC 1131-3 standard. Tools like those of the Open Development Kit can therefore only be ported to a particular family of controllers by adapting or extending their functionality.

*Principle 2:*
*Provide for methods or* ***options to adopt the system*** *to the users needs.*

Even manufacturer-independent tools like Open DK need manufacturer-specific code generators and ON-LINE drivers to adapt them to suit the conditions in existing systems. Since this customizing process always affects the manufacturers' own interfaces, porting is only practicable with their co-operation. The advantage is that the independent tools are exactly matched to the control system and perfect functioning is guaranteed by the manufacturer.

## 3 SUPPORTING THE REUSE OF EXISTING SOLUTIONS

In the past, the various controller families from each manufacturer all had their own programming units with specially developed programming tools. Changing to a different controller meant investing in new software development equipment each time and programmers had to spend their weekends poring over a new manual and getting to grips with new terminology and worse still a new programming language or at least a new dialect.

In switching over to a different supplier, the user has to rewrite large parts of the application. This and the fact that there is no common education in using these tools separated the industry even more.

> *Principle 3:*
>
> *Preserve* ***upward compatibility*** *with existing user source code. If this is not possible, supply tools for porting application software to the new environment without loss of information.*

The essence of mechanical or process engineering technology are the algorithms used for solving frequently recurring problems. To enable the engineer to employ the same method repeatedly within an application, IEC 1131-3 makes provision for the creation of instances.

The declaration of an instance of a function block must not be confused with the invocation of a function block. Creating an instance of a controller FB is the same as using several hardware modules of the same type in an automation solution.

Defining an instance is simply the creation of duplicates with identical functions. The introduction of the standard has vastly improved the conditions for extending the development of reusable application software to the PLC sphere.

> *Principle 4:*
>
> *Support at most the* ***reuseabilty of existing applications*** *to an extent as far as possible. The productivity of programmers depends to a large extent on the degree to which he has access to solutions already developed.*

A program organisation unit (POU) should be declared as a function block if it is a program part which is frequently needed and required to be reusable. An FB can have several input and output parameters and - unlike a function - can also store internal data. The values of the output parameters and internal variables are retained when an FB is called. For this reason, invocation of an FB with the same input parameters does not necessarily always yield the same output parameters.

Direct access to the inputs and outputs of a PLC is not possible within an FB. This makes FBs hardware-independent. Any FB that has already been declared can be used in the declaration of another FB or program.

An equally important feature compared to function blocks is the concept of functions! The first parameter of a function is the intermediate result of the current calculation and it always yields a new intermediate result. If the function has exactly one parameter, it cannot be distinguished in practice from an instruction because IEC 1131-3 does not stipulate parentheses for the parameters of functions.

This facility for apparently extending the instruction set of the controller is available not only to the manufacturer but also to the user. Therefore the reusability of solutions may be enhanced defining a functional substitute for a missing instructions on a target systems which doesn't support the full functionality despite the fact that there might be a slight loss in performance.

Clearly, the most important task of the Project Manager is to help the user manage configurations, resources, programs, function blocks and functions. But what about all the other information which is generated during editing, compiling, generation of target code and downloading?

We have the philosophy that the casual user should not be burdened with worrying about where to store those files. This task should be devoted to the Project Manager utility. On the other hand, the experienced user, who has several incompatible PLCs to program will use the same FBs and functions and as far as possible, only generating code for different targets out of the same sources. This user needs a sophisticated tool to support his task.

Furthermore with the emerging standard there will be numerous suppliers of function block libraries which are used as predefined functional units already tested and approved. Using not only its own source but also the geniality of numerous engineers will greatly enhance the productivity.

> *Principle 5:*
>
> *Provide for the possibility of **third party FunctionBlock-Library** support.*

The project manager is a central graphical tool which provides an extremely user-friendly, efficient interface for managing the programs, functions and function blocks of a project. Using the project manager the user can call individual blocks for editing, print them or transfer them to the controller. It also assumes central control of access to these structures by other tools.

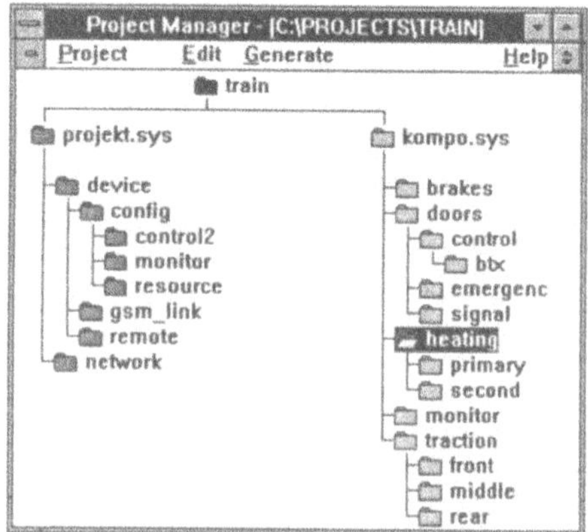


Figure 2: Enhancing reusability through a Project Manager

One of the major tasks of the Project Manager is to support the user by structuring his project and to manage program organization units in an efficient way. Project data can be accessed via an open interface, which can also be used by other programs.

The project manager is used for creating and structuring projects. The actions performed with the project manager bear an outward resemblance to actions which can also be performed at operating system level or using a file management program. When you create a project containing branches, a directory with sub-directories is created.

The entire compiling and linking procedure for a complete project can also be controlled using the project manager.

# 4 DESIGNING THE USER INTERFACE

Surely five programming languages are enough for one subject? Five languages probably, but five editors are definitely not sufficient. In fact, different types of applications call for different versions of the editors, tailored to the special requirements.

This is the reason why we offer two editors for FB-Programming: A configuration oriented CFC-editor and the FBD-editor which was constructed for highly interactive programming.

What is common to all implementations of interactive editors is that all are designed with the highest possible comfort for the user in mind. Comfort means that each task the user wants to perform has to be implemented in a way that user interactions are minimised as far as possible. To phrase it more generally:

> *Principle 6:*
>
> *Design for* ***easy modification of designs and programs*** *because 80-90% of tasks are changing already written programs and only 10-20% is writing new code or designing new programs.*

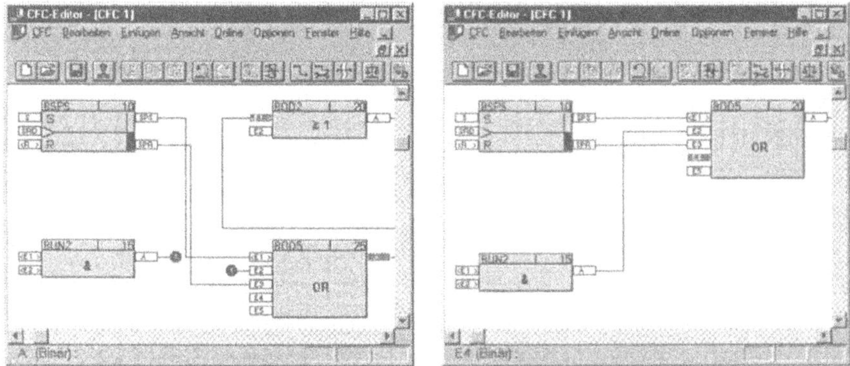


Figure 3: Moving function blocks with autoplacement and autorouting

The solution the developer provides to the end user can be easily measured: The cost function is defined as total number of keyboard hits and mouse movements necessary to perform a given set of alterations on an already existing program. The best implementation minimises this cost function.

Studies of PLC programs in industrial engineering have revealed that only between 10 and 20% of all control programs are developed from scratch. The majority (80 to 90%) are at most updated. This means that most of the time the editors of a programming system are not used for writing or developing programs, but merely for updating existing programs.

Just compare the systems currently available! How many user actions does it take to move the function in the example above to different position? Only three or four user actions are necessary. Very few of the CAD-oriented systems available on the market can manage with less than 20 or 30 actions. Unfortunately, it is not possible to assess the quality of a programming system at first glance.

Much harder to achieve and even more complicated to judge is the requirement of designing sophisticated engineering tools for the casual user or sometimes untrained personnel.

> *Principle 7:*
> *Design a system with a **predictable behaviour**. Assure the user that each action that he takes will be checked as soon as possible. Make sure that he can rely on the fact that the system will either accept his instructions or reject them immediately.*

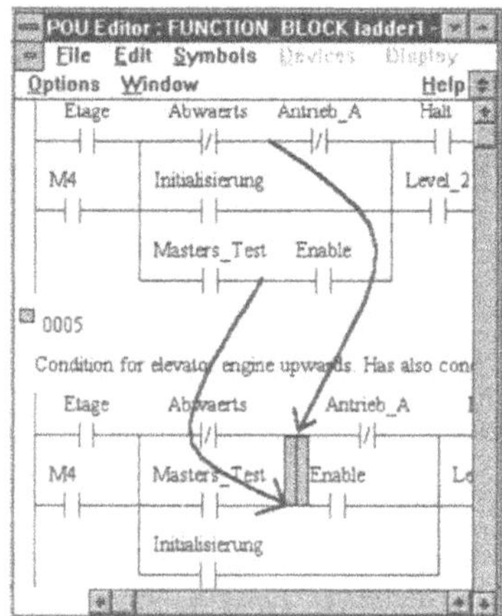


Figure 4: Automatic checked graphics in ladder diagrams

The method of graphical program development by means of repeated transformation of valid, correct graphical structures has been used successfully since the beginning of the 80s. Why should users settle for less today?

The graphical ladder diagram editor only accepts input operations which result in correct networks. User entries are immediately checked as to whether they can be represented in graphical form and unacceptable entries are rejected. This ensures that the current network is always graphically correct.

Programs and blocks entered in the graphical languages can also be displayed and tested in IL. Conversely, programs written in IL can also be represented in a graphical format as far as the graphical facilities allow. This fact is due to the incremental compiling method by which all actions of the user are compiled in transformations on a language-specific working set. The effect of this transformation is shown to the user by the visualisation module.

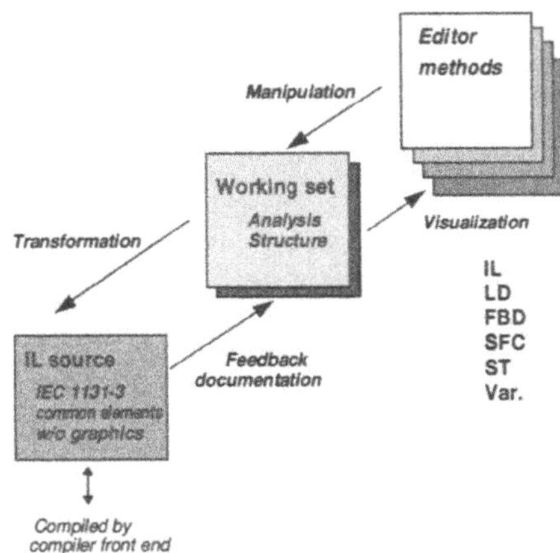


Figure 5: Graphical transformation using incremental compiling

The same goes for the introduction of the mouse for programming. Users demand from experience that all tool operations should be possible from the keyboard.

> *Principle 8:*
> *Design the system to be used with **keyboard-actions** and allow for mouse-control. Not vice versa! With the widespread use of windows oriented systems this feature is not fancy any more, nevertheless it is still important.*

As can be seen with the example of inserting shortage between two rungs, an incremental system does most of the graphical rearrangements by itself!

The graphical editors all employ the operator input concept: *Mark* (object or area) *and then perform the action.* This speeds up inputs and alterations considerably and also enables inputs to be made exclusively on the keyboard as well as by using the mouse.

When marking program parts it is possible in both graphical and textual editors to select single objects, whole areas or the entire program. In the graphical editors, insert operations with objects selected in this way are always subjected to a syntax check.

For single marked graphical objects the method of pre-defined editor actions is used, whereby certain default actions - e.g. double click or space bar - trigger standard operations on the object concerned. These graphical elements with attributes make it easy, for example, to negate or insert a contact or change it into a labelling field.

Full support is provided for the editing functions of the standard Edit Menu (Cut, Copy, Paste, Find, Replace) and the use of the clipboard. For the graphical languages the correctness of the syntax of the program is guaranteed in operations using the clipboard.

All users of interactive systems know that a little bit of uncertainty or selecting the wrong option like „cut“ instead of „copy“ can make the work of some hours all in vain. Modern systems therefore provide sophisticated aids for undoing and redoing actions to recover from the mess we may create under pressure (like writing this article late in the evening).

> *Principle 9:*
> ***An untrained user fears that an action he takes will lead to disastrous effects. Make sure that each action that has a chance of loss of work already done has to be checked or at least can be made undone.***

## 5 ON-SITE DEBUGGING MADE EASY

For testing and start-up, programs and data need to be exchanged between the controller and individual tools. The function of the communication manager is to present a universal view of the controller, i.e. to conceal the actual PLC from the tools. This enables the tools to be manufacturer-independent and ensures an open and efficiently adaptable interface to the controller.

These logical connections are used, for example, for transferring current I/O values via the on-line interface from the controller to the editors or for loading the PLC program into the controller for remote control. Depending on the functionality of the connected remote device there exist several features for remote control of a PLC:

- Remote control (start, stop, warm restart)
- PLC status display: RUN, STOP
- Program name and versions
- Program comparison PC <> PLC
- Status information
- Monitoring of inputs/outputs
- Force/Set via direct I/O

Using simulation it is possible to run the program in different cycles and to debug the program by a stepwise execution without any hardware attached. This gives a powerful possibility for „black-box-testing" the function blocks and the program.

A more typical example is the execution of an application while the process and the controller is up and running. Monitoring of parameters of a Function block in a graphical environment can be done in real-time without disturbing the performance of the PLC. Some purists argue that this is not true engineering style but such is life!

> *Principle 10:*
> *A typical control-application is not designed, it is engineered. That means it is programmed on the plane to the facility where it will be installed. Because real application development is done by trial and error, a programming system must support* ***experimental programming with continuous improvement.***

To support the user in this engineering task we have to provide tools which allow a better recognition of binary transient signals and have the possibility to have a powerflow display of binary variables. This is already known from debugging ladder diagrams and a proven technology which greatly simplifies the recognition of what's going on with the process.

It is important to note that such features could not be substituted by pure monitoring of global variables, because the parameters depend both on the calling environment (the program which calls the FB) and the actual instance which is called (the duplicate of a FB which has a unique internal data storage):

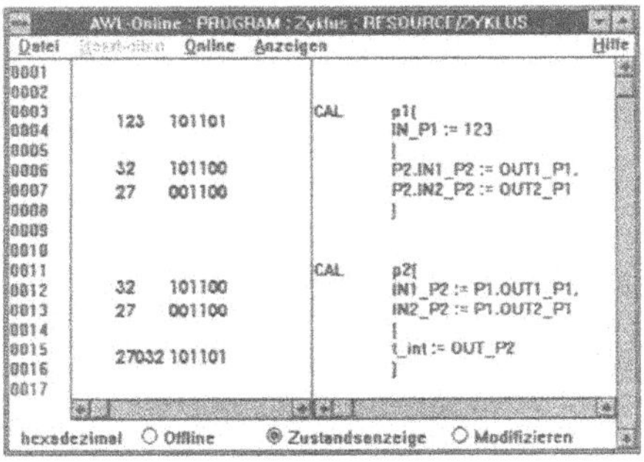

Figure 6: Online monitoring instances of function blocks in a running process

## 6 DOCUMENTATION OF APPLICATIONS

A PLC is part of the machinery in which it is integrated, it is not a computer. Because a machine lives from 5 to 10 years, the chance that it has to be modified by someone else, who has no floppy disk with the original program - or at least no computer which can read in those old-fashioned floppies.

The features of the documentation system play a major role in selecting an appropriate system from the engineer's point of view. That is because listings are the only documentation the end user gets.

Therefore sophisticated documentation features are a must and include:

- Complete program listings in graphics for programs, function blocks and functions
- Sub-projects and the project tree, cross reference list
- Declarations, instructions, assignment list
- Compiler results, program comparison results
- Controller configurations, on-line statuses etc.

*Principle 11:*
*Provide for* ***re-documentation of the application*** *solely from the PLC.*

When it comes to documentation of the application the printed documents are intended to be used by personnel who don't know much about the algorithms the implementers used and in most cases are unfamiliar with the programming environment. So if the documentation contains terms or expressions which are unknown to them the documentation may be of no use at all.

*Principle 12:*
*Use* ***terms and verbs*** *which are convenient for the user. Best is to use the vocabulary he or she already knows.*

Because real programmers don't read manuals this applies also to the help files, menus, dialogues and other stuff!

## 7 CONCLUSIONS

This paper shows principles which should be obeyed by every designer of CASE-Tools for control automation. In applying these principles to modern programming environments infoteam Software created a software package which is widely accepted because of its reliability and its user-friendly interfaces and methodologies.

## 8 REFERENCES

Brendel, W. (1994) Objektorientierte Werkzeuge zur SPS-Programmierung. *Automation Precision 10/94, 10-12*

Brendel, W.; John, K.-H.; Grötsch, E. (1994) BASE LEVEL Certification of IEC 1131-3 compliant products. *PLCopening April 1994, 6-7*

# 9 BIOGRAPHY

Dr. Wolfgang Brendel studied computer science at the „Friedrich Alexander Universität Erlangen - Nürnberg“ in the early seventies. Special emphasis was on graph-grammars and incremental compilers. He then joined Siemens and was responsible for the development of the STEP 5 programming software tools until 1981. After rejoining the university he worked on hardware description languages and silicon compilers finishing with the degree of Ph.D.

In 1983 Dr. Brendel, with three colleagues, founded „infoteam Software“, a company in the software services industry located near Nuremberg, Germany. Nowadays infoteam is one of the leading manufacturers of PLC programming systems with regard to the new standard IEC 1131-3 and is working in a close cooperation with almost all renowned PLC manufacturers. Developing quality software is a main focus.

Infoteam is a company owned by the employees, which ensures a high degree of commitment between employee and employer.

# VPLC - A Case Tool for the Virtual Programming, Simulation and Diagnosis of PLC-Software

*o. Prof. Dr.-Ing. Dieter Spath*
*Dipl.-Ing. Peter Guinand (guinand@wbkst19.mach.uni-karlsruhe.de)*
*Dipl.-Ing. Marco Lanza (marco.lanza@mach.uni-karlsruhe.de)*
*Dipl.-Ing. Ulf Osmers (osmers@wbkst4.mach.uni-karlsruhe.de)*

*Institute for Machine Tools and Production Science*
*Kaiserstraße 12*
*76128 Karlsruhe, Germany*
*Tel: ++49 / 721 / 608-4011*
*Fax: ++49 / 721 / 699153*

**Abstract**

Software development for programmable logical controllers is usually based on low-level languages such as the instruction list or the ladder diagram. At the same time, the programmer looks at a machine or an assembly system in a bit-oriented way; he translates the operational sequences into logical and/or time based combinations of binary signals described by means of Boolean algebra. This classical method causes a lot of problems in reality so it should be improved. It is the aim of the report to show a way developing PLC-software graphically and interactively within a Virtual Reality (VR) based system (VPLC).

**Keywords**

Computer Aided Manufacturing (CAM), Virtual Reality (VR), Programmable Logical Controllers (PLC)

# 1. INTRODUCTION

## 1.1 Outlining the problem

The process chain for the planning of PLC-controlled facilities gains more and more importance in companies of the mechanical engineering industry. However, in a company with an organizational division of tasks according to function, processes which actually intertwine are subdivided into partial processes that are often worked upon by „widely spread" specialized departments (Ritter-90). Therefore, information and communication problems typical for process chains (as shown in figure 1) occur here, too

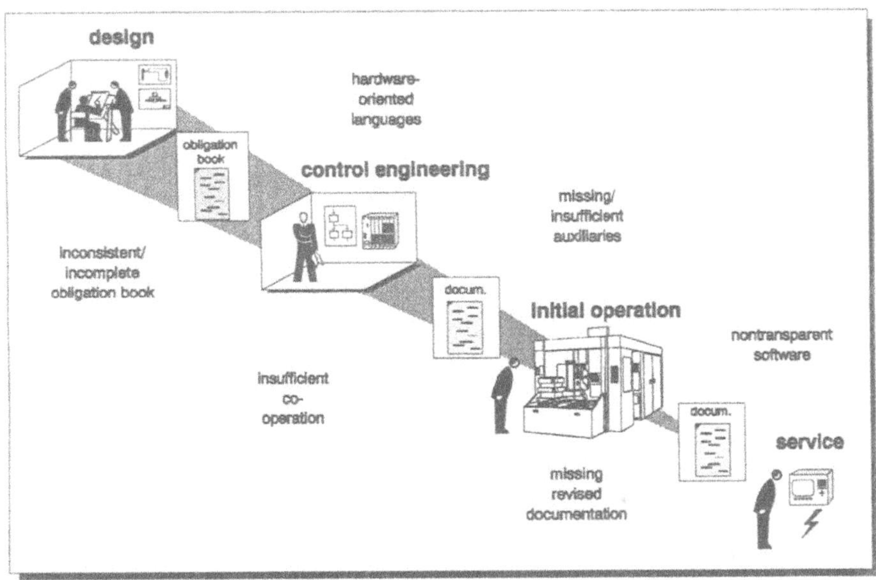


*figure 1: Problems of the process chain for the planning and maintenance of PLC-Software in companies of the mechanical engineering industry*

A detailed analysis showed that the lack of a uniform and consistent consideration of the facility to be controlled emerges as one of the major reasons for software errors. To-date, any properties and features of the facility of functional relevance have to be determined from different sources (CAD drawings, part lists, component catalogues, verbal/formal functional descriptions), the processed and integrated into the software by the control technician. Simultaneously, the efficiency of the development and test tools is highly limited due to the classical connection-oriented programming methods (Schelberg-94). Thus, at the interface between design and control engineering, a fracture in model making occurs which encompasses today the splitting up and at least temporary loss of information that was originally connected (figure 2).

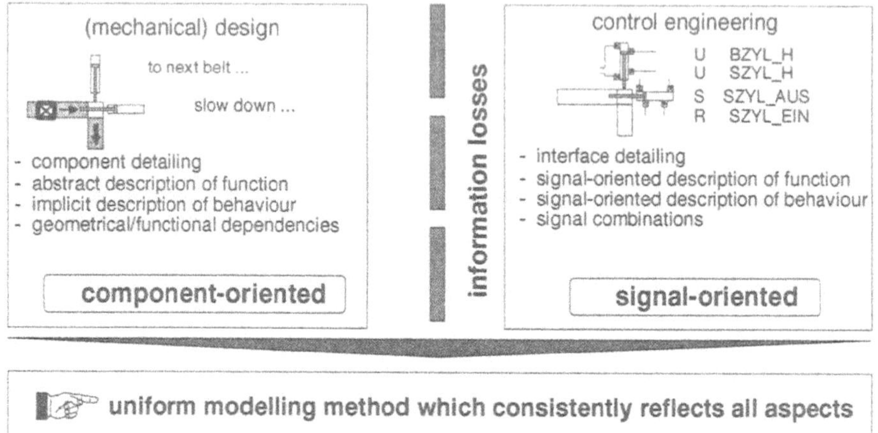


***figure 2:*** *Fracture in model making between design and control engineering*

One of the decisive stages of realizing automated facilities is the stage of initial operation since only here the orderly interaction of the mechanical, electric, hydraulic/pneumatic functions and those of control engineering of a facility can be checked. Many errors, the reasons of which can be found in prior stages, are recognized only during the stage of initial operation. Apart from high costs they may also result in safety risks for man and machine.

Therefore an important work basis for the staff in commencement of operations - who are usually not involved in the planning phase but are often supported by staff members from design and control engineering - is a sufficient description of the facility (Lanza-95) Apart from the data provided in the technical description, a three-dimensional visualization of the facility to be assembled can prove to be a great support for the assembly staff.

Different research studies have shown that particularly control engineering and there again the software development can be held responsible for a major part of errors occurring during this stage (figure 3).

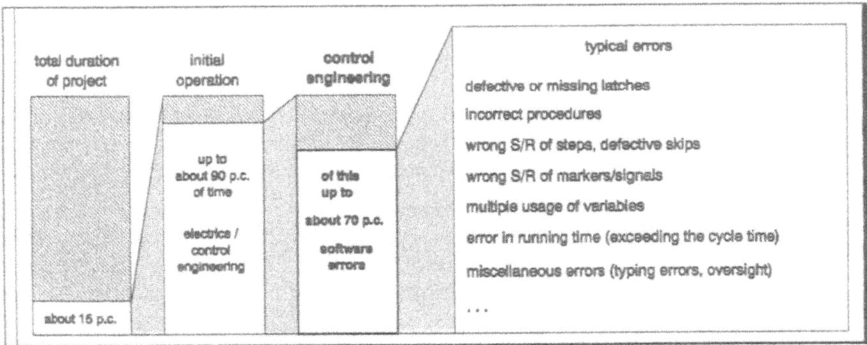


*figure 3: Research results of analyses of the initiation of operations*

## 1.2 Solution approach

A solution of the above named problems in the planning of automated facilities, namely

- sources of documents are distributed and sections overlap
- information losses between mechanical engineering and control engineering
- staff in initiation of operations have no spatial and functional comprehension of the facility
- software errors which cause great losses in time during initiation of operations
- no real graphic support, therefore no program evaluation in the development platform respectively in the projection stage
- complicated maintenance

is offered in the form of a uniform, integrated, computer-aided three-dimensional modeling of the complete facility including the implicit controlling task. However, the increasing complexity of the tasks in design, planning and production means that even the currently used methods of communication between man and computer have met with their boundaries.

Efforts to cope with this condition have led to the now extensive introduction of graphic user interfaces. These may mostly be used with the help of pointing appliances and they require far fewer abstraction skills and less knowledge from the user's part than the previous line or command-oriented surfaces did since they mostly use easy-to-remember graphic symbols instead of abstract commands. However, there are limits for the usage of these user interfaces as they are limited to two dimensions. It is particularly difficult to realize the handling of three-dimensional objects and working on problems which require a spatial representation.

New courses might be pursued by utilizing three-dimensional user interfaces the way they are realized in Virtual Reality systems. By using special input and output devices and computer platforms, these systems are capable of integrating the user or several users working in an interdisciplinary way (e.g. designer, layout planner, control technician) as the acting part into three-dimensional, synthetic environments which places the communication of man/machine and man/man onto a completely new basis (Osmers-95).

With VR systems, much more use can be made of the natural problem solving attitude of man than was possible with previous user interfaces since it is always possible to refer to real conditions due to the real-life arrangement of synthetic environments and the possibility of 3D-interaction in real time. The user is therefore able to act intuitively or based on his experience and can thus also treat more complex spatial problems without being impeded by limitations.

This view complies particularly with the designer who assigns technical basic components or assembly groups to the partial procedures of the functional structure and then composes the facility layout. Each assembly group possesses not only its geometry but also determined physical properties, a logical or dynamic behavior and last not least relationships to other components (Guinand-95).

The relationships and dependencies between the components may then be explicitly expressed by stating the relationship type; cardinalities (assessment of the relationships) are expressed by complementary data.

With the help of a function model, the basic procedures and connections within the VR system are described by the designer on a comparatively abstract level. Thus, on the one hand it represents the basis of an „intelligent" facility model together with the process model which may be regarded as a detailed and extended version of the function model, and on the other hand it is the basis of the subsequent PLC-program.

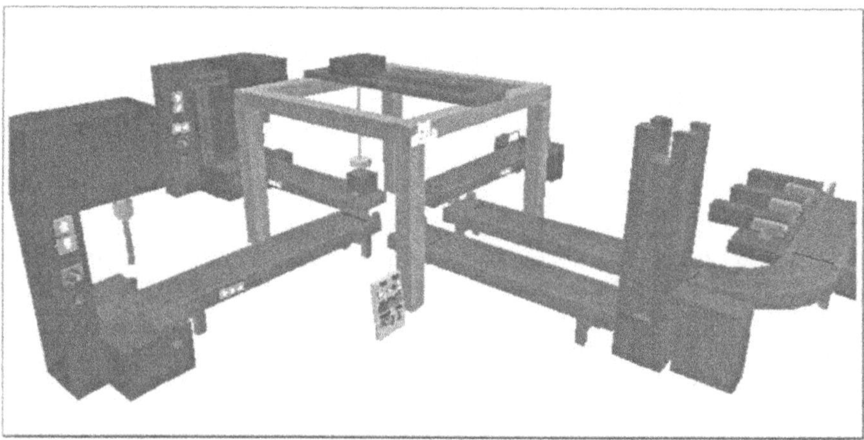

*figure 3: Exemplary VR-model of the facility*

At the moment, modeling is done with a commercially available VR-development environment and for the later stage a CAD-attachment is planned for the geometry and an attachment to an object-oriented data base for technological data.

## 2. REPRESENTATION OF THE OBJECT PROPERTIES

The systematic classification of the components of a facility shows that certain characteristics are common to several or to all of the assembly groups considered. All facility objects, for example, have a setup structure and a defined geometry while stating a logical behavior is generally only possible for functional units or assembly groups with their own information processing or control objects. Figure 5 gives an overview of the attributes and properties required for the definition in control engineering of a technical object. The modeling method allows any desired expansions at a later time (Spath-95).

### 2.1 Semantic data

This serves to manage the components. An essential item is the name of the component which will later facilitate unambiguous identification of the component type in any application. In this context it may prove to be an advantage across different company departments as regards the unambiguity of name to utilize the manufacturer's type name - e.g. that of a supplier of pneumatic components.

### 2.2 Geometrical data

With increasing computer performance it will be possible to import complex components from existing data structures or from widely used CAD systems without the abstraction required so far.

### 2.3 dynamic properties and attributing

Apart from the static (geometry) properties the dynamic ones of the facility elements have to be incorporated into the model.

An object may be structured using several dynamic elements. Their behavior and interaction are described and controlled by means of a program.

Path-time or velocity-time diagrams used to be the standard to describe the conditions of movement and the movement profile. Attributing within the VR systems allows the description of direct physical properties such as initial velocities, friction, pulses, kinematics chains, or modified movement profiles as can be observed with pneumatic components. Therefore, collision studies can easily be carried out.

### 2.4 link interface

Every object is equipped with specific dialogues which are activated at running time together with the object. They serve on the one hand to facilitate communication between facility components and controlling, on the other hand also the exchange of messages of (intelligent) peripheral assembly groups among each other.

## 2.5 Multi media interface

To improve the handling of complex object and data structures, VR systems are equipped with additional functions attractive to the sensory tract of man, such as texture mapping and sound.

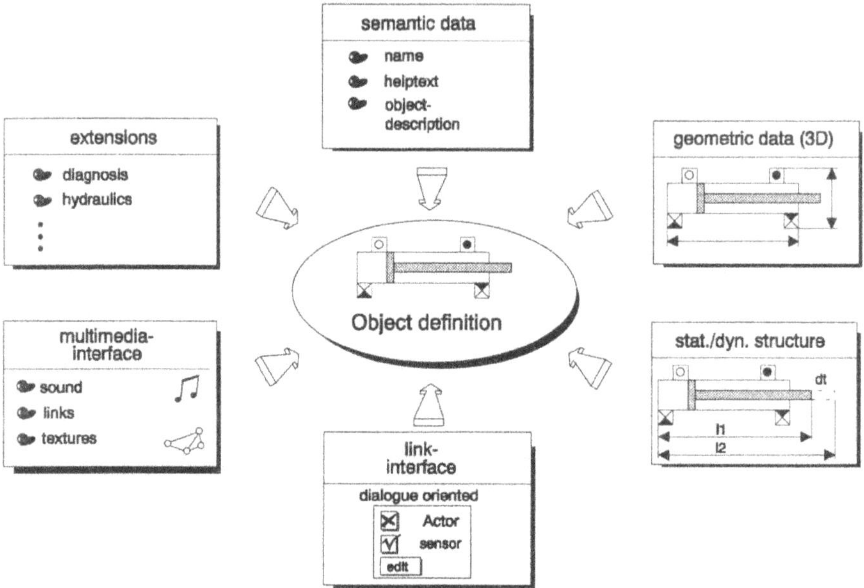


*figure 5: Definition of a data record to describe an object*

# 3. FROM THE OBJECTS TO THE PLC PROGRAM READY TO RUN

The following partial steps and task areas may be differentiated when planning VR-aided PLC programs.

First, the facility is supposed to be compiled with the help of working materials or machine catalogues. In the next step the logical connections between actor and sensors are produced by graphic, interactive programming technologies. Prior to the final realization, the configuration of the components (geometrical aspects) as well as the functionality of the procedure control (logic and time-related dependencies) can be validated and optimized in simulation runs. The final program is transferred directly to the real facility via suitable interfaces.

## 3.1 Configuration of the facility components

Using the working materials library and choosing from suitable menues (figure 6), the designer can load his facility modules, such as functional units (conveyer belts, pushers, linear axles), sensors (light barriers, limit switches) or even complete machines into his digital environment by clicking the mouse to „drag and drop". Apart from the geometry, the components are also furnished with certain basic functionalities corresponding to the modeling of chapter 2.

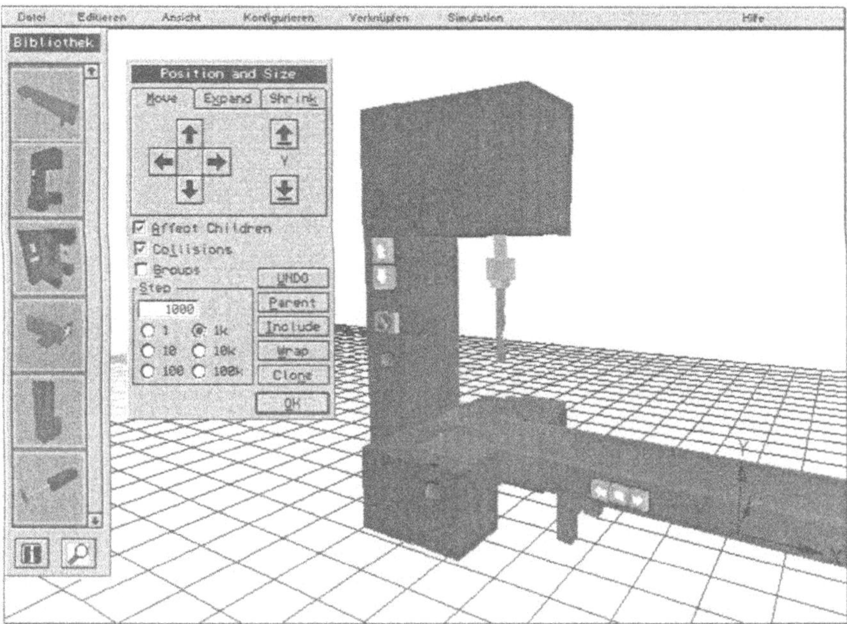


***figure 6:*** *Configuration of the facility layout*

With the help of a corresponding input device, such as a spacemouse (6 degrees of freedom), the user may freely navigate within the facility and can thus validate the layout. With the corresponding tools the facility layout can be adapted to different requirements at any time.

## 3.2 Logical-functional connection

Inlets and outlets (actors and sensors) are connected to become action modules by choosing easy-to-remember graphic representations of actions which reflect a „1 to 1" image of the mechanical elements in the real world.

Choosing the actions activates a dialogue which facilitates a „bonding" of the modules in the VR system in a graphic, interactive way with a low level of abstraction. Any complicated construction of relationships and restrictions in instructions list, ladder diagram or contact plan is redundant.

On confirmation the dialogue is valid and represents a functional module; the graphic representations are colored to mark „connection“.

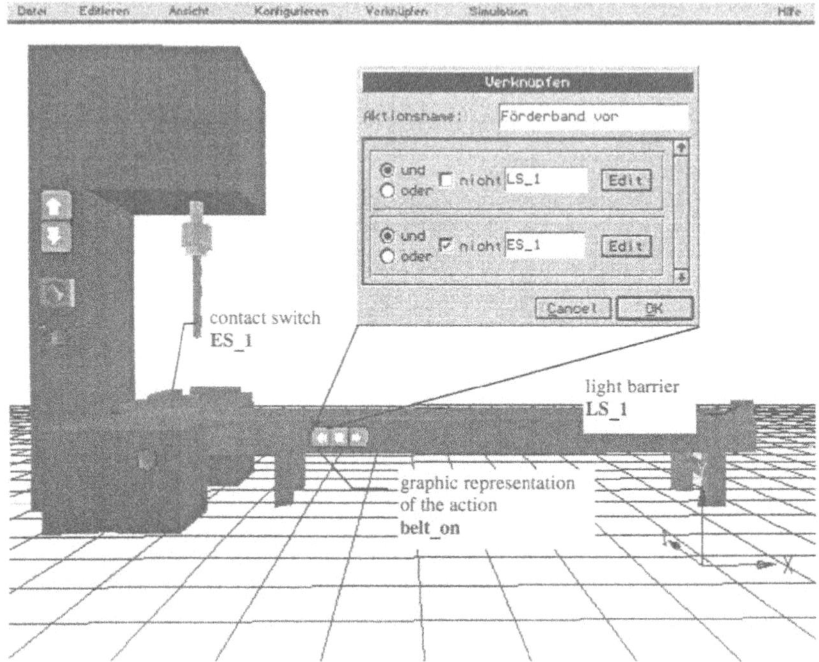


***figure** 7: Dialogue-supported programming of functional modules*

The logic of the functional module and thus even at an early stage parts of the PLC-program may be tested by simulating the partial procedure.

In the next step, the functional modules are connected to become procedure modules, again with dialogue assistance. If arranged in sequence, these modules show the procedure of the facility. This hierarchically structured „bottom-up“ approach is depicted in figure 8.

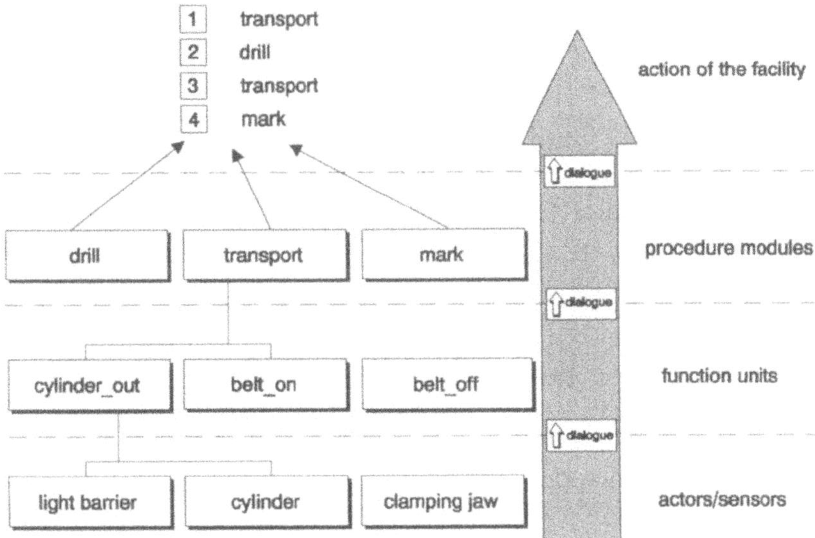


**figure 8.** *Aggregating the complete facility from the modules*

Upon completion of the programming, the facility is simulated taking into account aspects of time and geometry (collision). Errors, inconsistencies and weak spots are quickly recognized and eliminated due to the provided options of graphic representations, interaction and navigation within the digital model. Partial steps may be tested at any time, e.g. by resetting the tool manually in VR.

In the background, suitable transfer records according to DIN IEC 1131 (IEC-93) are issued. They can be translated into the different PLC dialects depending on the employed controlling. Apart from the control engineering, the layout plan, component and parts list may also be generated from the digital model.

In contrast to high-end graphic VR systems, the development on a PC-based system is advisable here since the focus of the application is not so much the graphic representation but the interaction and the advanced man-machine-interface. Moreover we expect greater acceptance in industry for the PC-based system.

## 4. OUTLOOK

The project VPLC is currently being realized at the Institute for Machine Tools an Production Science at the Technical University of Karlsruhe, Germany. With the integrating platform and interface VR, not only the pure planning of the facility and the PLC-program generating but also planning of hydraulic systems, monitoring, diagnosis and failure detection based on an intelligent model could be implemented.

## 5. REFERENCES

(Guinand-95) Guinand, P., Planning of production sites with the help of Virtual Reality Exemplary Realization, diploma thesis, Universität (TH) Karlsruhe, 1995

(Schelberg-94) Schelberg, H.-J., Objectoriented Planning of PLC-Software, Dissertation Universität (TH) Karlsruhe, faculty of mechanical engineering, 1994

(IEC-93) IEC1131 Part 3: Standard for programmable controllers - Programming Languages, Beuth Verlag 1993

(Lanza-95) Lanza, M., Schellberg, H.-J., Object Oriented Planning of control techniques, proceedings of „4. Fachtagung zum Entwurf komplexer Automatisierungssysteme", Braunschweig , June 7-9. 1995, p.463-468

(Osmers-95) Osmers, U., Visualising, animating, simulating - Virtual Reality makes your plannings alive., IHK-Informationen für die Wirtschaft, 9/95, p.17-19

(Ritter-90) Ritter, K.H.: process chains in product development, VDI-Breichte Nr.830, 1990.

(Spath-95) Spath, D., Osmers, U., Weber, J., Planning Production Systems with VR, DFG-Schwerpunktsprogramm, Ergebnisbericht der 1. Antragsphase, July 1995

## 6. BIOGRAPHY

o.Prof. Dr.-Ing. Dieter Spath, born in 1952. Study of mechnical engineering at the Technical University of Munich. In 1981, a doctor's degree at the Institute for Machine Tools and Management Science at the TU Munich. Entering the company group KASTO in 1981, he has been managing director of the same since 1988. In 1992 appointment to the post of professor in ordinary at the University (TH) of Karlsruhe, Institute for Machine Tools and Production Science.

Dipl.-Ing Peter Guinand, born in 1966. Study of mechanical engineering at University (TH) of Karlsruhe. In 1995 diploma degree and since 1995 working as a free reasearcher for the Institute for Machine Tools and Production Science (wbk) and at the Institute for Industrial Building Production (ifib). His main field of interest is the development of VR applications within the scope of CAD/CAM.

Dipl.-Ing Marco Lanza, born in 1966. Study of Production Science at the University (TH) of karlsruhe. In 1994 diploma degree and then research assistant at the Institute for machine tools and production science in Karlsruhe. His main fields of interest are lower cost automization and applications of Engineering Data Management Systems (EDMS).

Dipl.-Ing Ulf Osmers, born in 1967. Study of Production Science at the University (TH) of karlsruhe. In 1993 diploma degree and since 1994 research assistant at the Institute for Machine Tools and Production Science in Karlsruhe. His main fields of interest are the applications of Virtual Reality in Product Development and Production Planning as well as distributed manufacturing and scheduling.

9

# ASPECT - a CASE-Tool for Control Functions Originating from Mechanical Layout

*T. Brandl, R. Lutz, J. Reichenbächer*
*Institut für Steuerungstechnik der Werkzeugmaschinen und Fertigungseinrichtungen, Universität Stuttgart*
*Seidenstr. 36*
*70174 Stuttgart*
*Germany*
*Tel.: ++49-711-121-2420*
*FAX: ++49-711-121-2413*
*thomas.brandl@isw.uni-stuttgart.de*
*rainer.lutz@isw.uni-stuttgart.de*

**Abstract**

The systematic design of control software requires exchange of information between the software development department and other departments of an enterprise (e.g. project planning and mechanical design). For this purpose neither construction plans nor program listings are suitable description forms. In this paper three design levels and the related description forms are introduced. Thus a common information base shall be provided to all experts who are involved in an automation project. Step by step this information base may be detailed until it is possible to generate the code for a controller by a compiler automatically. Special respect is given on the reuse of design objects. For the efficient application of the proposed method the CASE-Tool prototype ASPECT has been developed at the Institute for Control Technology at the University of Stuttgart.

**Keywords**

Control software, software design methods, reuse, CASE-Tool

# 1 INTRODUCTION

The present development of control software is characterized by increasingly complex requirements in regard to functionality, cost and quality. Control software, once only an "accessory" of electrical machine design, has now become an indispensable part of the final product, the "machine".

In the past the essential task was the programming of binary logic as a substitute for an relais control. Now, various additional tasks are to be solved, as for example diagnosis and monitoring functions, communication, man machine interfaces and data processing like tool management or machine data acquisition. These requirements cannot be met with traditional approaches and new techniques are required. This paper will show how control software development can be improved and which methods and tools are needed for this improvement.

# 2 PROBLEMS AND GOALS

The current process for developing control software is characterized by a strictly sequential succession of machine design tasks (Pritschow, u.a., 1994). Little exchange of information happens between the departments involved (Fig. 1). A lot of information (which is important later for the programming) is generated at an early phase of the project e.g. as a requirement specification or a submission to the customer made by the project planning department.

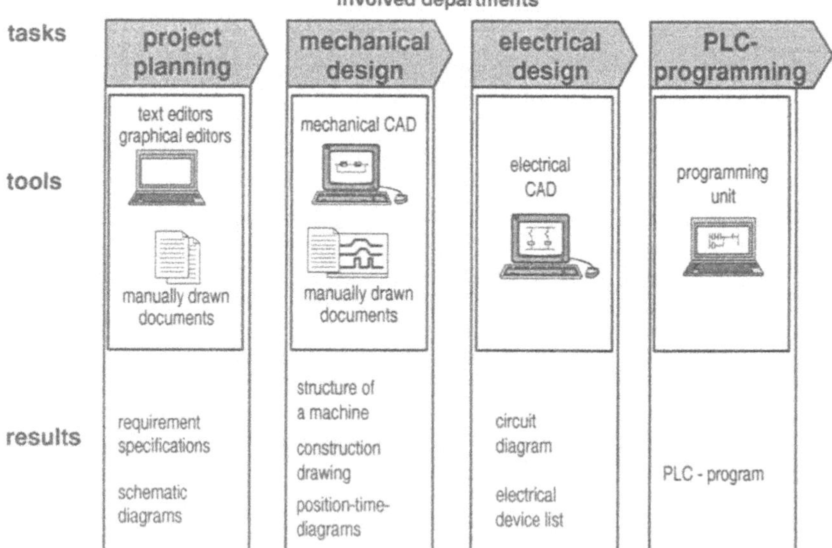


**Figure 1** Typical sequence of project processing. (104 339e)

Usually, computer-aided tools are applied, as e.g. mechanical or electrical CAD. Since it is impossible to exchange logical data between the different tools through suitable interfaces, each tool represents an isolated solution in regard to a specific area. Consequently, a great amount of data has to be entered repeatedly which results in increased work and a higher error rate.

In addition, these data isolated solutions impede an efficient cross-departmental application of re-usable design solutions and machine modifications. Therefore, reuse is only possible in specific departments; there it is executed at a growing rate. But this means that a minimal modification made in the mechanical design department may result in considerable modifications and adaptations of the control program.

In regard to cross-departmental information exchange, the question always arises: Which presentations and describing forms are most apt? It is very important to create a **common communication basis** (Storr u.a., 1994) for the staff involved. Not every document currently used is suitable for this purpose: a construction engineer will not know how to use an instruction list program, just like an PLC programmer will not derive benefits from the details of a technical drawing.

The customer driven necessity to change from one controller vendor to the product of another one usually requires a re-programming. The reuse of already developed and tested program modules is not possible in this case. Even the standardization of programming languages according to IEC 1131-3 (N.N., IEC 1131-3) did not improve this, because this standard does not include portability specifications.

Along with the actual development, the preparation of the technical documentation determines the elapsed time of a project. Information from the individual departments are usually collected and made into a document after the technical development is finished. The primary purpose of this is to make comprehensive information available to the service and maintenance staff. An accurate preparation of the documents is also important in regard to product liability or CE certification.

The above mentioned problems could be solved by the following measures:

**Improving software quality** by

- continuously proceeding from the start of the project to the commissioning,
- application of widely understood methods for the design of control software,
- increased re-use of already generated software design and modules and
- developing device-independent designs and programs.

**Shortening** of the elapsed **project time** by

- parallel project development by means of early information transfer between departments,
- preparing the documentation along with the development,
- automated generation of program code and
- application of one or more suitable computer-aided tools.

## 3 REDUCED TIME FOR CONTROL SOFTWARE DEVELOPMENT

Reduced development time for control software must allow in particular for the transfer of information at an early stage (Weck, Kohring, 1991). Here it is useful to provide different categories and to assign these categories specific functions. An extensive analysis leads to three design levels (Storr, u.a., 1994; Herrscher, Grimm, Storr, Reichenbächer, 1991)

- general description (design level 1)
- functional description (design level 2) and
- detailed description (design level 3).

Fig. 2 shows which departments in the different levels generate the descriptions and which departments use them.

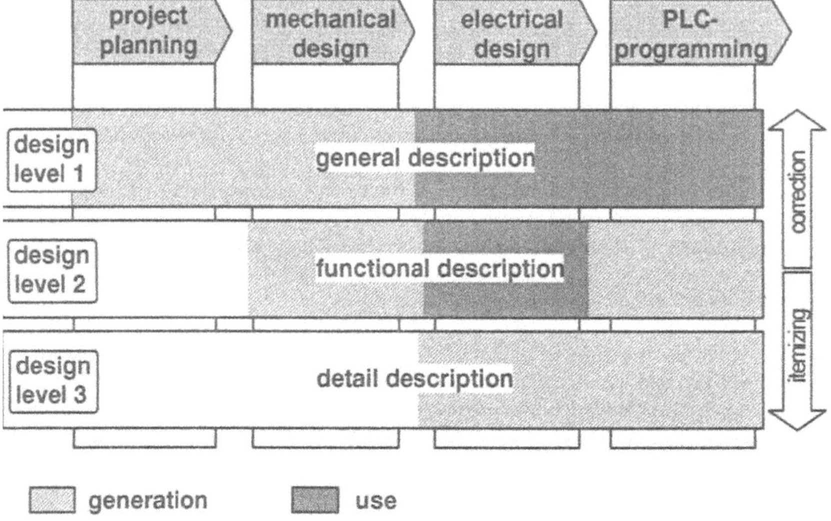


**Figure 2** Design levels and description forms (104 340e)

**General descriptions** are mainly used for the early structuring of a machine or system, e.g. during the project planning phase or in the beginning of the mechanical design. Functions and mutual dependencies are illustrated in an abstract form.

The purpose of **functional descriptions** is to break down the functions that first, in the general descriptions, have been displayed abstractly, in order to fulfill the technical requirements.

**Detailed descriptions** give more information about all additional attributes of the elements used in the other design levels, besides just structure and function. Such information may be measurements, identifiers, cross references, connected loads etc.

By introducing defined structuring levels, we facilitate the modularization of the entire machine function and functional descriptions and design methods for control software. Thus the functions to be implemented can be described long before technical solutions are detailed. So it is possible

- to provide functional information at an early time and
- cross-departmentally aim at applying standardized solutions.

Early data transfer is, at the same time, a necessary premise for parallel representation of the project work (simultaneous engineering). Appropriate methods for describing the functionally oriented structure and for the software design are presented in the following chapter.

# 4 GENERAL DESCRIPTIONS AND FUNCTIONAL DESCRIPTIONS

Suitable methods for supporting a parallel and cross-departmental information exchange in machine tool builder enterprises require a view of the problem that differs from the generally accepted views of today which are specific to particular departments. This means, for example, that functionality of a control system is not necessarily understood as a PLC-program for a specific controller. First there has to be a rather exact design for the control system, in which the functionality required (and later to be implemented) can be recognized by other participants of the project.

The layout of an installation is determined by mechanical design engineers. Therefore, it is convenient if the design objects which the control software engineer uses are closely related to the design objects of the mechanical engineer. Only if both experts agree for which parts of an installation already existing design objects are to be applied; can efficient reuse be possible.

At the design level of general descriptions, an **expanded technology scheme** can be used for this purpose, which differentiates design objects and contains a coarse geometric representation (Fig. 3). Design objects incorporate function units, function groups, actuators, sensors, operator control elements etc. as well as the interdependencies between them. Function units and function groups (in the following called function objects) can be detailed or combined so that any complex structure can be built. The design engineer can specify the behaviour of each function object by means of functional descriptions. The definition of clearly separated design objects results in

- unified structuring of a machine or a system,
- simple generation of modules and consequently easier reuse,
- clear function specification
- possibility to describe mutual dependencies and relations and
- high comprehensibility.

As a next step, by using the descriptions of design level 2, it is necessary to add formal definitions to the design objects. This applies, as a rule, to function units and groups. Design objects like actuators and sensors are not further detailed. The relationships between the function units should also be described, since they represent the data exchange through interfaces of the function units.

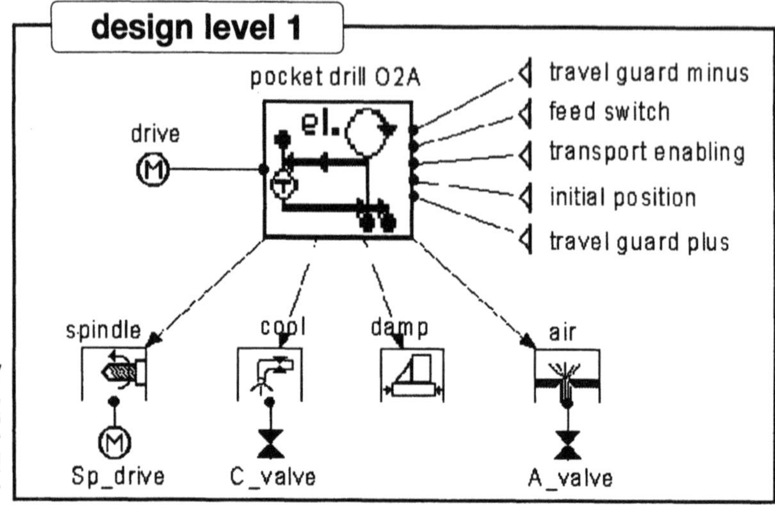


**Figure 3** Example of an expanded technology scheme (104 341e)

Suitable descriptions (Pritschow, u.a., 1988) for logical dependencies - especially because of the graphical representation - include sequential function charts, Petri nets and state graphs. Signal curves according to VDI 3260 (VDI-Richtlinie 3260, 1977) describe more time-oriented interrelations.

**State graphs** (Fig. 4) (Fleckenstein, 1987; Otto, 1992) are examined closer in this book in the paper „State Diagrams - A New Programming Method for Programmable Logic Controllers". They can be used for describing function units and function groups formally and up to the code generation stage. State graphs are intentionally device-independent and easy to understand in order to enable different departments of an enterprise to use them. Nevertheless, they can be used to directly generate control programs - a feature that is required for continuity. For this purpose the description of all design objects contained in the general description must be completed. In detail this means

- complete state graphs for all function units and groups and
- sufficiently detailed specifications of the components (addresses, data types, etc.)

Thus the code generation for any type of controller is possible and also easy to automate after the details have been specified in design level 3. The I/O-addresses of controller signals are such a detail, for example. Fig. 5 shows our concept and examples of the code generation process. We have found that certain features of a language like subprograms, case differentiation constructors and symbolic identifiers are important for a direct mapping of the design objects onto the target language. In the case of control-oriented languages like the "Modicon State Language" or structured text according IEC 1131-3 this is exceedingly simple to do. Assembler-like languages which include in essence all current controller dialects as well as three languages of the IEC 1131-3, require much more effort. Fig. 4 illustrates a simple realization with function blocks where efficiency of the code depends on how much the target control supports the processing of function blocks. The application of special compilers might therefore be worthwhile.

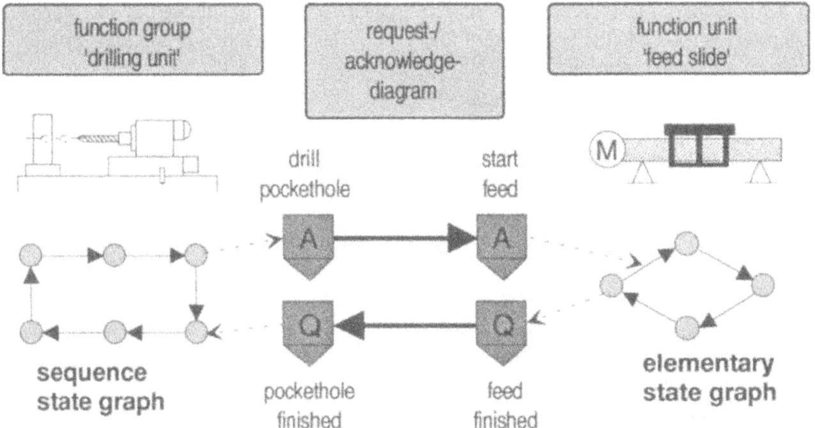


**Figure 4** Sequence and elementary state graph synchronized by requests (A) and acknowledgements (Q) (104 343e)

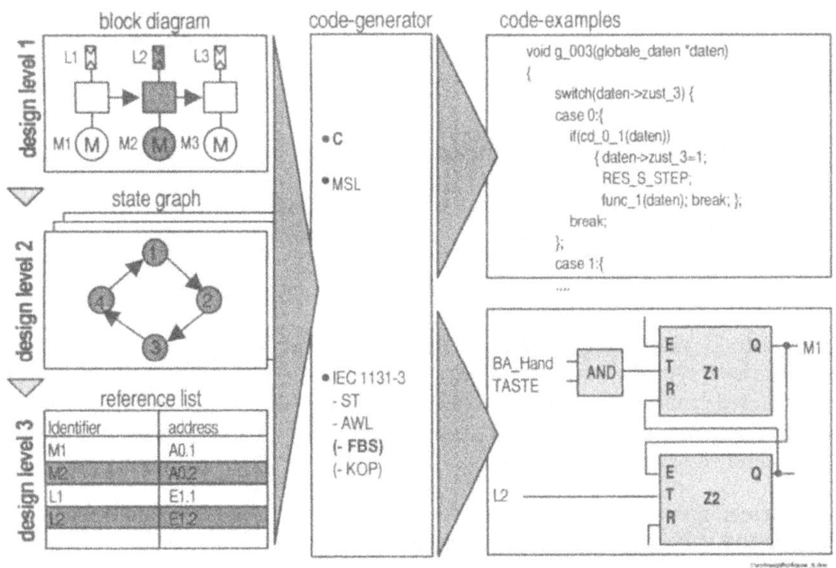


**Figure 5** Concept of code generation (104 344e)

## 5 CASE-TOOL ASPECT FOR SUPPORT OF CROSS-DEPARTMENTAL CONTROL PROGRAM DEVELOPMENT

To a large part the methods and descriptions mentioned in chapter 3 and 4 meet the requirements placed in the beginning of this paper. The introduction of a computer-aided tool promises additional benefits (Lutz, 1995) because

- the employee is relieved of routine work,
- data is kept consistent and
- the current isolation of today's CAD systems can be partially abolished.

Important requirements for such a tool include

- support for all three design levels (general diagrams, functional descriptions, detailed descriptions),
- facilitates continuous progress throughout the project,
- uniform and consistent administration of entered data,
- simple and intuitive user interface,
- configurability by the user,
- extensibility,
- broad applicability, and
- possibility for multiple access.

Fig. 6 shows the structure of the prototype developed at the ISW, which fulfills the mentioned requirements for the most part. In particular the integrated project data base ensures relevance and consistency of data. The unified structuring allows the integration of further function modules in ASPECT.

## 6 CURRENT RESEARCH ON REUSE

Our current research work attempts to increase the efficiency and quality of software by specific and systematic reuse of designs, based on the state graph method. Reusable design provides the following benefits:

- Lower maintenance time and cost because only a few modifications have to be maintained even if a large number of applications are in operation.
- Less faulty and more stable designs, because faults are detected early due to repeated application of control software modules.
- Improved productivity because the know-how of qualified personnel is incorporated.
- Good and well established design examples make for an additional training effect that is especially beneficial for new personnel.

The present object-based modeling concept and the support by ASPECT create a basis for a systematical software development. Design levels and description methods provide the possibility to copy complex design objects which are congruent to mechanical subsystems from existing applications and to reuse them as software design modules.

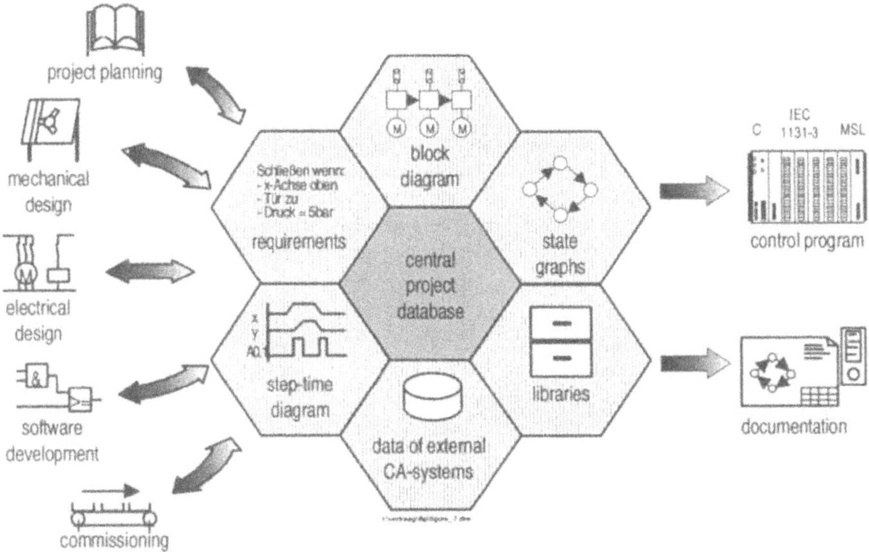


**Figure 6** Modules of the CASE-Tool ASPECT (104 346e)

But systematic reuse for practical applications exceeds the „copying from application A and inserting in application B". It requires a defined procedure with additional descriptions. So software development becomes a systematic configuration with parameterizable modules which are administrated in libraries. This results in the following features (see also Fig. 7):

- For the administration in libraries the function objects have to contain information details (reuse information) in addition to the software design. Thus a successful search is possible in an abstract manner, without knowing the design details. Such information includes the specification on design level 1, list of hardware equipment (actuators and sensors) and the abstract service interface presented to the outside.
- Function objects have to be configured for the application when extracted from the library. Configurable elements are stored within the components as place holders. Place holders include identifiers (of design objects, states, components, variables,..) or comments. For the assignment of values or texts to place holders, function objects have to be represented by corresponding templates.
- By connecting the interface symbols (orders and acknowledgments) the function objects are integrated in an existing application. Since these interface symbols postulate specified counterparts (corresponding client and server functionality), they can, under certain circumstances, be searched for and connected by computer. So the configuration is partially automated.
- Objects of all design levels and of any complexity have to be specified and administered in the library. This includes simple binary sensors as well as complex hierarchical function groups (e.g. drilling units) with subordinated function units and their sensors, actuators and

function descriptions (e.g. sequence and elementary state graphs). Furthermore, we need to examine how subfunctions (e.g. subgraph 'motor acceleration') can be defined and administered for repeated application.

- In order to classify modifications to function objects, the object-oriented principle of 'inheritance' has to be applied within a class hierarchy. Through inheritance, variable declarations (data) and functions can be generally used for similar function objects and redundancies can be avoided. In contrast to textual object-oriented languages (e.g. Smalltalk), a graphical language for control applications requires additional mechanisms, which allow for the inheritance of data (variables within function objects), hardware (sensors and actuators with controller inputs and outputs) as well as graphical design descriptions (e.g. state graphs).

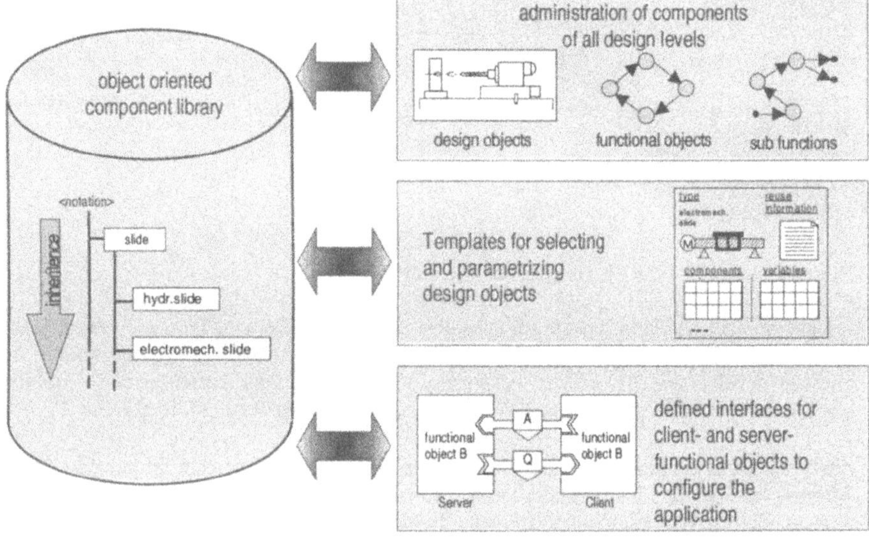


**Figure 7** Systematically reuse of software design (104 345e)

The implementation of the above mentioned features is the subject of a current research project sponsored by the German Ministry for Education and Research, which is conducted in cooperation with two mechanical engineering firms and two software enterprises.

## 7 SUMMARY

With the current approaches to the development of machines and systems, the weak points of software engineering have been demonstrated. Strictly sequential procedures and poor information exchange between the departments of an enterprise are typical. An improvement can be achieved by a parallel procedure, the use of widely understandable description methods, a higher degree of reuse of existing design objects and the use of supporting tools.

The development process has been divided into three design levels "general description", "functional description" and "detailed description". The usage of appropriate description forms allows for an improved project development time. The descriptions contain all information which is needed to generate the control code automatically by means of suitable compilers. The development engineers are therefore no longer compelled to take care of control hardware related details in early phases of a project.

This procedure is efficiently supported by a modular and open CASE tool, which makes the development of control functions easier through the use of graphical editors and also integrates all departments involved in the project through a central project data base. For this purpose the ISW developed the tool ASPECT as a prototype.

Current research work at the institute aims in particular to further improve the support for modeling and reuse of machine control software, by using object-oriented techniques.

## 6 REFERENCES

Storr, A., u.a.(1994) Simultan zur SPS-Software - Neue Ansätze zur effizienten SPS-Programmierung. Elektronik 23, 124–136.

N.N., IEC 1131-3: Programmable Control, Part 3: Programming Languages.

Herrscher, A., Grimm, W., Storr, A., Reichenbächer, J. (1991) Systematische Software-erstellung für Steuerungen. In: Tagungsband zum FTK '91, Stuttgart 1.–2.10.1991, 52–58. Springer-Verlag, Berlin, Heidelberg, New York, Tokyo.

Pritschow, G., u.a. (1988) Studie über mögliche Beschreibungsformen bei der Software-erstellung und Dokumentation. VDW-Forschungsbericht 1010, Frankfurt.

Fleckenstein, J. (1987) Zustandsgraphen für SPS - Grafikunterstützte Programmierung und steuerungsunabhängige Darstellung. ISW63. Springer Verlag, Berlin, Heidelberg, New York, Tokyo.

Otto, H.-P. (1992) Zustandsgraphen - Eine Einführung in die Methode mit Beispielen. Druckschrift des Arbeitskreises "Lastenheft für ein SPS-CASE-Tool", Nürnberg.

VDI-Richtlinie 3260 (1977) Funktionsdiagramme von Arbeitsmaschinen und Fertigungsanlagen. Beuth-Verlag, Berlin, Köln.

Pritschow, G., u.a. (1994) Interdisciplinary Models and Descriptions for the Program Development for PLCs. Production Engineering Vol. II/1. Hanser Verlag, München, Wien, New York.

Weck, M., Kohring, A. (1991) Die Bedeutung der Anlagespezifikation für die Entwicklung von SPS-Software. In: Pritschow, G., Spur, G., Weck, M. (Hrsg.): Maschinennahe Steuerungstechnik in der Fertigung. Hanser Verlag, München, Wien, New York.

Lutz, R. (1995) Systematische Softwareerstellung im Maschinenbau mit dem CASE-Tool ASPECT. In: Software-Entwicklung: Methoden, Werkzeuge, Erfahrungen '95; 6. Kolloquium 12.–14. September 1995, Technische Akademie Esslingen, Ostfildern.

# 7 BIOGRAPHY

Dipl.-Ing. T. Brandl has worked at the ISW as electrical engineer since 1991. He carried out various research and development projects in the area of control technology, mainly for PLC-controlled special purpose machines. Currently he is developing a tool to present the technical documentation as an „electronic manual". Since 1996 he has been head of the group „Software Engineering and Diagnosis" at the ISW

Dipl.-Ing. R. Lutz After receiving his degree in mechanical engineering at the University of Stuttgart in 1993, he began to work at the ISW in the group 'Software Engineering and Diagnosis'. His main research field is methods and tools for software engineering in control technology and he has made a major contribution to the design and development of the CASE tool ASPECT.
Within the joint research project MOWIMA, which is sponsored by the German Federal Ministry of Education and Research, he is currently working on the reusability of software by developing and using module libraries for machine and plant construction.

Dipl.-Ing. J. Reichenbächer started his career at the ISW in 1988 and was head of the group „Software Engineering and Diagnosis" from 1990 to 1995. One of the projects he managed during this time was the development of the CASE-Tool prototype ASPECT that is presented in this paper. In January 1996 he left the institute to build up a group for development of master control technology at Charmilles, which is a manufacturer of erosion machines.

10

# Case Tools for Flexible Manufacturing Systems

*Prof. Dr.-Ing. Dr.-Ing. E.h. M. Weck, Dipl.-Ing. J. Friedrich, Dipl.-Ing. Th. Koch, Dipl.-Ing. R. Langen*
*Laboratory for Machine Tools and Production Engineering (WZL), Aachen University of Technology, Germany*

**Abstract**

Case Tools for application engineering in the field of flexible manufacturing systems (FMS) have gained importance due to the rising complexity of the control software. This paper presents three different case tools from the current research activities at WZL which are part of COSMOS, an open control architecture for flexible manufacturing systems. *CellDesign* is a graphical development tool for cell controllers using a petri net based approach. *CASCADE* provides a framework with class libraries and design patterns for an object oriented development of shop floor applications. *MMS-3D KIT* is a set of different Case Tools for the development of device drivers which are needed to integrate machine tools incompatible with the Manufacturing Message Specification (MMS).

**Keywords**

Case Tool, Cell Control, Class Library, Design Patterns, Development Environment, MMS, Open Architecture, OSI, Petri-Nets, Shop Floor Control, Software Engineering, Workframe

## 1 INTRODUCTION

Enterprises, large and small, are more and more organized into manufacturing cells to simplify business control, enable lean production and increase the efficiency of production life cycle. These changes require new control systems based on new technologies such as open systems, client server architectures and object orientation which offer flexible and efficient solutions. The major hurdles in entering this new world are related to software: the time to develop it, the ability to maintain and enhance it, the limits of a program's complexity with regard to its maintainability and the time it takes to become familiar. This leads to the major issue information systems have today: long time to market, insufficient quality, high cost and lack of interoperability. While hardware cost are decreasing, software expenses are still rising.

In this paper new and powerful Case Tools used to develop software for the COSMOS control architecture for flexible manufacturing systems (FMS) are presented (Figure 1). COSMOS is intended to provide efficient, coherent and cost effective control of manufacturing process at the factory level (Weck 1995) by providing a generic and open control architecture. It is based on an integrating infrastructure linking the diverse control applications into an integrated, but distributed system.

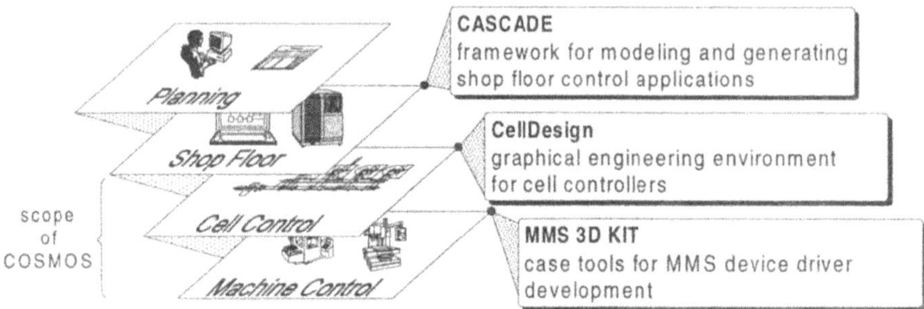


**Figure 1** Case Tools of COSMOS.

Due to the complexity of flexible manufacturing systems the development of control software requires a systematic engineering approach to improve software quality and reduce development time and cost. Therefore, user oriented software engineering tools are an essential part of the COSMOS architecture, e.g.:

- **CASCADE** is a development environment consisting of several tools for modeling and generating *shop floor control applications*. The tools of CASCADE support an object oriented approach which is based on two different types of objects: resource objects and dynamics objects. The application framework provides base classes and design patterns for these objects.
- **CellDesign** is a graphical engineering environment for *cell controllers*. It combines the power of petri nets with the benefits of reusable software. While basic functions of a cell controller are provided by so called Function Objects, applications are designed by glueing these functions together graphically.
- **MMS „3D“ KIT** are diverse Case Tools needed for the *MMS device driver* development. The MMS 3D-Kit enables the integration of machine tools using communication standards other than the International Standard Manufacturing Message Specification.

COSMOS provides a powerful software architecture that enables FMS control systems to be defined, developed and installed in the minimum time and with the maximum possible reuse of existing components.

# 2 CELLDESIGN - A CASE-TOOL FOR CELL CONTROLLERS

Due to the complexity of Cell Controllers software development requires a systematic engineering approach to improve software quality and to reduce development time and costs. Though many CASE tools are available today they are often not suitable for the development of cell control software. This is because they either do not provide powerful reuse concepts or they do not support modeling of event-driven systems with lots of parallel execution, but interdependent processes (Frey, 1992). For that purpose a specific CASE tool called *CellDesign* has been developed at WZL.

## 2.1 Basic Concepts of CellDesign

The client-server concept of COSMOS provides a mechanism for the development of cell control applications based on reusable software. It is suggested that control applications may only contain the control flow of a program while the function execution itself is dedicated to servers. A function call in an application causes a message to be sent to the corresponding server (Figure 2). This approach has two major advantages. Firstly, application modeling is independent from function implementation. Secondly, services (functions) provided by existing servers are reusable. As a consequence powerful and generic server libraries help to increase the reusability and to decrease the implementation effort.

In cell control applications sequential data processing, algorithms and control of dynamic processes can be distinguished. For modeling of transaction oriented and algorithmic applications, methods such as flow-charts are available. Unfortunately these methods are not suitable for the design of parallel, dynamic and asynchronous processes, which are characteristic of manufacturing cells. Therefore a modeling approach based on so called *action-nets* has been developed. The basic idea is that applications are modeled as objects which react to external events. For example, an event called "new cell order" is sent by a shop floor control system to the cell controller, as shown in Figure 3 . Such an event is specified by

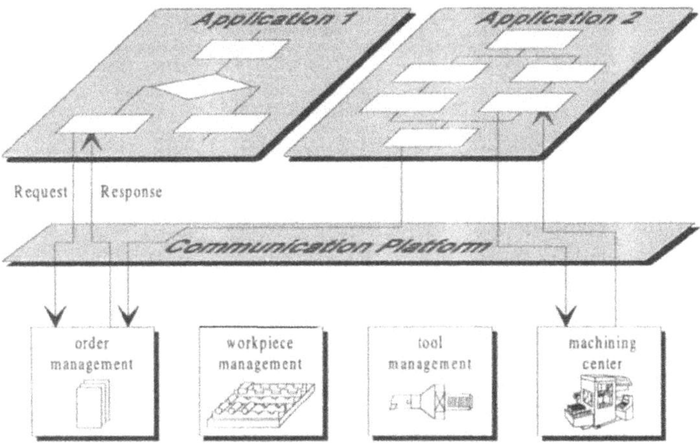


**Figure 2** Separation of control flow and function examination.

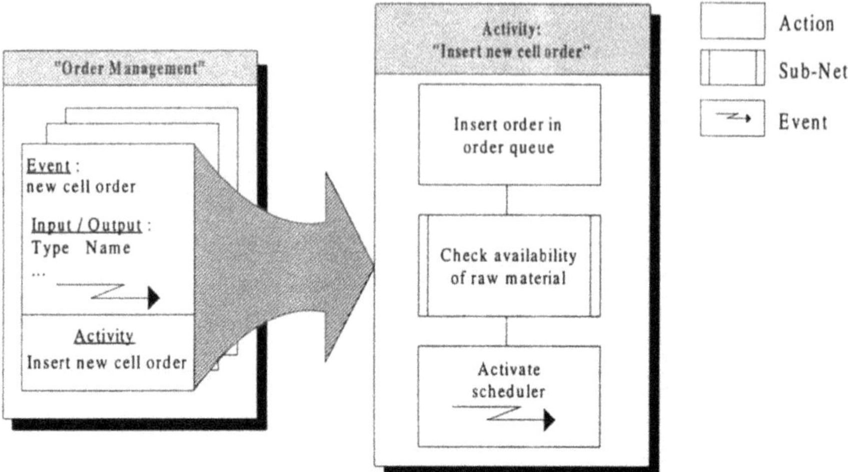


**Figure 3** Events, activities and action nets.

a unique identifier and a list of input-data. An event causes an *activity* of the application which can be described in more detail by a sequence of *actions*. Actions are either atomic functions executed by servers or further activities in the sense of "sub-nets".

Modeling of control structures in action-nets is based on mechanisms provided by petri-nets.

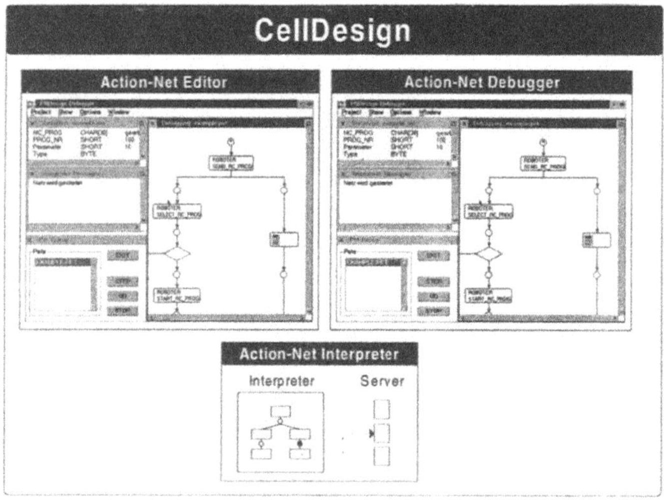


**Figure 4** Integrated Development Environment of CellDesign.

A state of a petri-net can be considered as a precondition for the execution of an action which is represented by a transition. An action will be executed only if all preconditions are fulfilled respectively marked. With petri-nets it is very easy to model control structures like sequence, loop or parallelity and they also simplify synchronization between parallel processes.

Based on these concepts for application engineering a CASE tool called *CellDesign* has been developed at WZL, which supports the overall process of application design and development. As shown in Figure 4, this CASE-Tool consists of three components: action-net editor, action-net debugger and action-net interpreter. The editor is a powerful, fully graphical oriented tool for action-net design. It provides all basic elements like states, actions and connections as well as elements which represent start and end of an action-net.

At the end of the design phase, the code generation process must be activated. This process compiles all action-nets into a formal language, which can be interpreted by the run-time system or by the debugger. For debugging the graphical net representation will be used to ensure that a system developer is able to test an application at the same level of abstraction and representation as used in the design phase.

The execution process of an action-net is represented by marks flowing through the net. In "step-by-step" mode the execution of action-nets is under full control of the developer. Furthermore it is possible to monitor all variables of an action net.

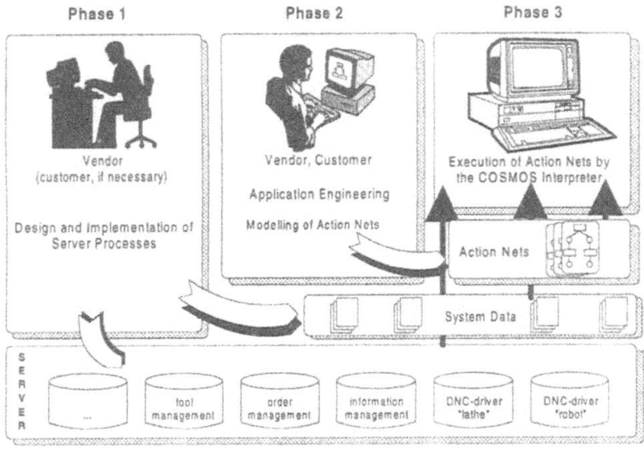


**Figure 5** Phases of Application Engineering in CellDesign.

Figure 5 summarizes the engineering approach for the development of end-user specific cell controllers based on reusable software. In phase I software vendors define basic and reusable functions combined in servers which represent the "core functionality" of a cell controller.

In phase II the action-net editor will be used for end-user specific application design on top of the existing "core functionality". Functions which are not available have to be implemented separately. At the end of phase II the code for application execution will be generated automatically, so that there is no more implementation effort for the developer in phase III.

# 3 CASCADE - A CASE-TOOL FOR SHOP FLOOR CONTROL APPLICATIONS

Looking at recent developments in shop floor control systems there is an increasing demand for systems offering a higher degree of adaptability. The first generation of shop floor control systems had a monolithic structure (Kohen, 1986) which was superseded by a second generation of systems having a modular architecture (Lange, 1993, Pritschow, 1991). The adaptability of these systems is based on the exchange of entire modules. However the possibilities concerning adaptation do still not readily facilitate individual solutions. Object-oriented technologies offer promising possibilities.

According to Zipper (1994) ease of use, flexibility and the capability of being integrated are the most important requirements for shop floor control applications. These features are necessary to guarantee long-term use. Therefore the following requirements on the design of a class library for shop floor control applications should be fulfilled:

- separate modeling of the dynamics processes and the classes itself
- low degree of links between classes
- simple applicability
- extendibility / maintainability
- flexibility / universality
- stability

The application framework developed for CASCADE predefines the architecture of the software system by fixing the structure and interaction of the single components (Booch, 1994, Pree, 1994). This framework approach does not only support reuse of software at the source-code level and the domain specific class hierarchy. In addition it supports the reuse of the entire system-architecture. Therefore a rudimentary, application independent executable skeleton is available. It can be adapted to individual demands by modifying the application specific „hot spots“ (Pree, 1994). For this purpose the user is supported by a graphical tool set as described above. This allows a high degree of reuse, because not only the class library (based on design patterns) (Pree, 1994, Gamma, 1994) can be reused, but also the „glue“ between the components, (e.g. exception handling).

## 3.1 Modeling approach

The modeling approach supported by CASCADE, a toolset for modeling and generating shop floor control applications, is based on two types of objects, resource objects and dynamic objects. Real objects physically existing in the production system are modeled as resource objects - sequences and strategies of material flow etc. as dynamic objects. Resource objects describe objects such as machine tools, transportation units, buffers, tools to be supported by the software system. They are modeled applying the notation as presented by Rumbaugh (1991), and consist of attributes, methods and a state model. Instead of identifying an object's state by one variable (LatheState = {WORKING, DEFECT, ...}) the state can be described by any combination of the object's attributes. Dynamic objects incorporate the interaction of resource objects. Processes within the manufacturing system are described in this type of object. Dynamic objects are modeled using scenarios describing these processes. The reason for introducing two types of objects is that manufacturing systems differ mainly in the processes to be carried out within the system. Therefore primary modifications for adaptation can be carried out in the dynamic objects.

## 3.2 Development environment

CASCADE is a development environment consisting of several tools for supporting a user during the phases of modeling and application generation by applying the above approach. A *repository* is the central integrating unit for the tools, serving as the storage for data generated during the design and development phases. It is implemented using an object-oriented database on an OS/2 platform. Basic tools are the *Resource Object Builder* and the *Dynamics Object Builder* (Figure 6).

They are needed to build the model of the system according to the approach described above. The design and development cycle is stored in the repository. Afterwards a syntactic check can be carried out using the *Model-Verifier.* If the verification has worked successfully the software based on the model can be generated automatically using the *Application Generator.* A *Documentation Generator* enables the automatic extraction of information out of the repository and the generation of a paper-based documentation using a standard text processor.

The core of CASCADE is a class library which provides base classes and patterns to build resource and dynamic objects. The basic concepts of this library will be described next.

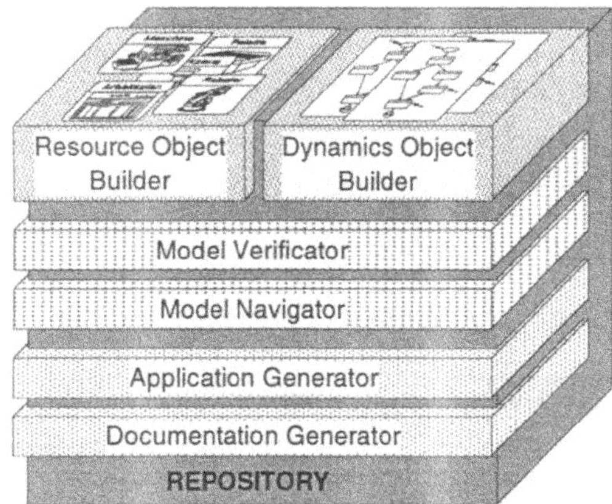


**Figure 6** Architecture of the Development Environment.

## 3.3 Basic concepts of the application framework

The class library designed for the development of shop floor control applications is implemented in C++ on an OS/2 platform and includes base classes with extensive features. Unlike other approaches for class libraries for shop floor control applications (Schmid, 1994) it includes fundamental concepts for software development. Therefore it not only provides classes for items like machine tools, buffers etc. The underlying concepts are important because they are the basis for the design and development of the entire class library.

Basic concepts include runtime type information (RTTI) (Chen, 1995; Stroustrup, 1992) as well as exception and contract handling. Other concepts which have been implemented are association, iterators, base classes for the application itself, the main program, dynamics objects, i/o-interfaces (e.g. terminals) etc. Furthermore, we distinguish between active and passive objects. Active objects have their own thread on a certain site e.g. an object representing a unit in the production system. Passive objects are used by active objects and represent items like tools and fixtures. Classes needed to give objects activity (including communication and distribution) have also been realized. The following paragraphs give a brief summary of some of the basic implemented concepts.

RTTI allows a dynamic (at runtime) type-checking in C++. Regular C++ only features static type-checking. RTTI is needed to avoid runtime errors using „containers", which are often used in shop floor control applications. Design by contract is used in several OOD-methods (e.g. Wirfs-Brock (1993)). This concept is also not supported by regular C++ (unlike Eiffel - see Meyer (1994)). To improve system stability and robustness this feature has been implemented in terms of precondition, postcondition and invariant. A precondition tests whether the client keeps the contract at the beginning of a method. A postcondition has to be valid after carrying out a method to enable a server to keep its contract. An invariant is a condition which should always be true. In combination with the implementation for the design by contract, an extended warning and error handling mechanism has been implemented using several error levels. An error carries out a *throw()* statement while a warning does not. Both statements put information in the logfile. However a program can continue its regular process after a warning occurred. An exception needs an error handler. For dynamic objects suitable „handlers" are available. For resource objects they have to be implemented by a developer depending on the application.

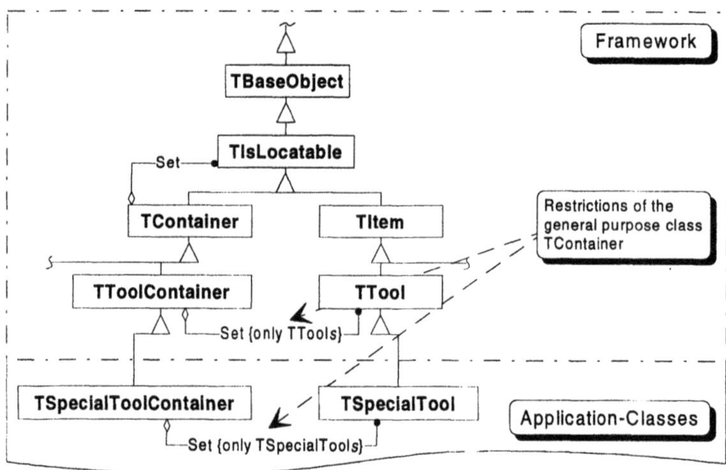


**Figure 7** Extract of the provided class library.

An association is one of three types of relations between classes and is used extensively in OO-methods. In shop floor control applications they are needed for example, to design the relationship between order and customer. An association could be implemented using pointers. However pointers are often used without modeling „real world" relationships. According to Rumbaugh (1991) associations should be implemented as bi-directional. For this reason a base class TIsAssociatable has been developed which allows one to build relationships between objects in the same sense as in the design.

To step through any type of container or list, a generic iterator concept has been implemented. Applying iterators allows uniform access to data, in addition to encapsulating the needed mechanisms and hiding them from the user.

Figure 7 shows a brief extract from the class library. The framework provides generic application classes like TToolContainer or TTool. TContainer and TItem are derived from TIsLocateable, which describes objects having an identifiable position within the system. TContainer itself can contain any type of locatable object. The class TToolContainer is derived from TContainer. However a restriction has been made on the types of objects this container can carry. Only objects belonging to the class TTool can be taken by instances of TToolContainer.

The mechanism needed to build this „special" container is shown in Figure 8. The method *Admit()* of the base class TContainer uses a precondition to test the object to be put into the container. The precondition uses the method *AdmissionTest()* to verify whether the object is of type TTool. This is carried out by applying RTTI (method *IsA()*).

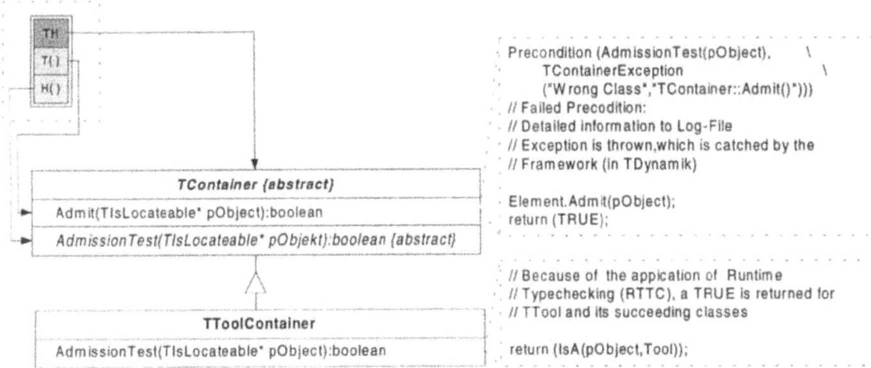


**Figure 8** Building a Special Container using the Base Class TContainer.

If a user wants to realize a restriction to TSpecialTool, the only modification needed is:
*boolean TSpecialToolContainer::AdmissionTest(TIsLocateable* pObject)*
*{ return(IsA(pObject, TSpecialTool)); }*

This brief example outlines some of the possibilities the class library offers to a developer. In a similar way patterns (e.g. for tool assembly plans, setting up a system layout etc.) have been implemented. These base classes enable a user to develop the resource classes in a natural way. In combination with the state model the needed flexibility for building resource objects is given. The dynamic behavior of the system can easily be changed by adapting dynamics objects.

## 4 THE MMS „3D" KIT: AN MMS BASED APPROACH FOR INTERACTIVE DEVICE DRIVER DEVELOPMENT

In order to accomplish the electronic information interchange between automation devices like machine tools, robots, etc. on the one side and control computers as well as DNC hosts on the other side, the control computers and DNC hosts have to be equipped with dedicated software modules. These software modules normally are called „device drivers". A device drivers task is to adapt the specific communication interfaces and communication behaviors of a specific automation device to the control computers or DNC hosts software.

Device drivers are commonly implemented as independent software modules having two different interfaces for the information exchange.

The first interface is the interface to the controlling and DNC manufacturing application software modules. To avoid interdependencies between the manufacturing application software and the device driver software modules and to avoid software customization, the device drivers need to be implemented in the same manner for all the different automation devices in the manufacturing environment.

The second interface of a device driver is used to exchange information with the remote controller of the automation device under control of the manufacturing application software. Here the information exchange is carried out via the different current device specific communication platforms.

For the various available automation devices nowadays the device drivers have to be specifically created. This is caused by the high number of different vendor specific communication protocols which were developed in parallel in the past as a result of missing communication standards.

This situation increases the implementation efforts and costs in a way which cannot be accepted by industrial system users. Therefore in recent times a large financial expenditure has been committed to the international standardization of vendor independent generic communication protocols for automation devices. Today as a result of these activities international standards are available which for example allow remote data base access (RDA), remote file handling (FTAM) and the remote control of automation devices (MMS with its associated Companion Standards).

The implementation of these standards in device controlers will decrease the development expenses because necessary software customizations in the manufacturing application software can be significantly reduced.

With MMS -that is the Manufacturing Message Specification- since 1990 an international standard is available solving the communication problems in manufacturing environments. But due to the current global situation in the metal-processing industry and due to the complexity of MMS controller software implementations, MMS communication interfaces have not found the wide spread use that they should have until now. Even today latest machine tool control developments are offered with a communication technology based on obsolete concepts.

But there is no serious doubt that MMS compliance will lead in the long term to automation devices with generic vendor independent communication interfaces (ESPRIT CCE-CNMA, 1995). Because of the above mentioned reasons and keeping in mind that most of the devices used in industrial manufacturing environments today are not equipped with MMS interfaces, at the Machine Tool Laboratory of Aachen University (WZL) a kit for graphical interactive device driver programming is under development (Figure 9). The kit is called the „MMS 3D kit" and is one of the results of WZL's research activities regarding generic device driver architectures. This architecture should allow one to create device drivers which fulfill the MMS requirements, which are independent of computer operating systems, which drive devices with vendor-specific communication interfaces and which provide a generic interface to the manufacturing application software.

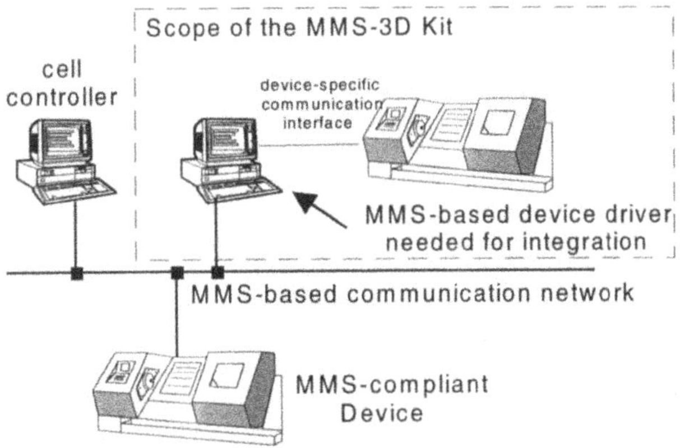


**Figure 9** Scope of the MMS-3D Kit

## 4.1 Kit Profile (Functionality)

### *4.1.1 Communication Interface between device driver and manufacturing application software*

In an object oriented way MMS has specified 11 different communication relevant object classes and 87 accompanying communication services. Generally, with a small subset of the provided MMS objects and services described in so-called MMS Companion Standards, the communication behavior of any automation device can be modeled in a generic way (Friedrich, 1992). Mainly the MMS object classes VMD (Virtual Manufacturing Device), Variable, Domain, Program Invocation and Event Management are needed.

The MMS 3D Kit provides modeling tools for the mapping of vendor-specific communication behavior to dedicated MMS communication objects and services. For this purpose the device drivers developed with the MMS-3D Kit apply the following MMS objects and services.

The VMD services Status, Identify and GetNameList are supported. The service Status allows one to check whether the device is ready for use or not. The Identify service makes it possible to get detailed information about the connected device. The names of all MMS objects implemented in the device driver can be requested with the GetNameList service.

The MMS-3D Kit permits mapping of Control-specific variables to MMS variables, which can be read or written with the corresponding MMS services. Furthermore the MMS-3D Kit permits the management of domains (e.g. NC/RC programs) within the device driver.

All Program Invocation services offered by MMS are supported by the MMS-3D Kit. In addition to that the device driver provides complete program management.

The device drivers permit the feature of the remote creating and deleting of event objects via a communication channel. With the support of these objects other MMS objects like variable objects can be monitored. If for example a user wants to be informed about the change in a specific variable value, the event objects pass the appropriate information to the MMS Information Report service.

### *4.1.2 Communication Interface between Device Driver and Automation Device*

Today different specific communication protocols are offered by almost any control vendor for the remote control of devices in manufacturing environments. The data packets are usually transmitted with line control procedures like the LSV/2 or 3964/R procedure which are often used as De-Facto standards in the German industry (Weck, 1995).

The MMS-3D Kit allows the modeling programming of the vendor-specific data packets described in the device controls operating manuals as well as the accompanying protocol. Even the programming of vendor-specific line control procedure is possible with tools from the MMS-3D Kit.

At the moment an open architecture for control systems is under development within ESPRIT project 6379/9115 „OSACA". A major output of this work is a communication platform for the interprocess communication within a heterogeneous control hardware is specified. To integrate OSACA control systems into an OSI environment, device drivers are required too. The task of these device drivers is the conversion of both MMS and OSACA protocols and vice versa. The MMS-3D Kit eases the programming of such OSACA-MMS device drivers.

## 4.2 Basic concepts

The device drivers developed with the MMS-3D Kit provide a generic, MMS conformant communication interface to the manufacturing application software. In the following the processing of an MMS service indication in a device driver is described (Figure 10). The data included in the service primitive of the service indication are temporarily stored in the global data area of the device driver. With the help of the MMS identifier received as information in the service indication, the corresponding MMS communication object is found in the object manager. The communication object that was found has a reference to an accompanying transaction net which is executed. The transaction net undertakes for example the task to send and to receive the packets needed for the communication with the automation device. The construction of the packets is carried out with the help of the global data. After receiving the response packet from the device control, a positive MMS response is issued. Otherwise the MMS response is negative.

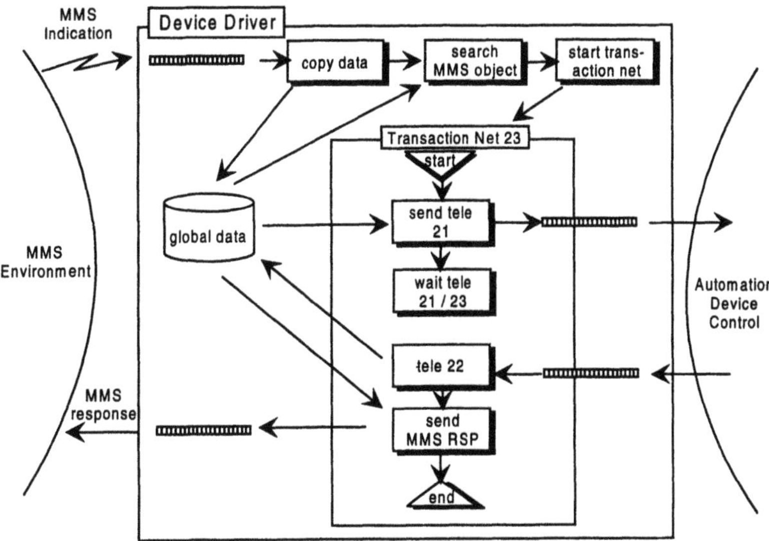


**Figure 10** Device Driver Architecture.

## 4.3 Kit Tools (Editors)

The previously introduced MMS indication processing within the device driver has to be programmed completely. It is the task of the MMS-3D Kit to reduce the actual programming work and the time needed for it and to reduce the programming faults to a minimum. The MMS-3D Kit provides the user six with tools for the development of the different steps needed for indication processing in a device driver. These tools are:

### *4.3.1 Protocol Editor*

The protocol editor is used for the graphic programming of line control procedures. The code programmed with the Protocol Editor is block oriented. Each block has clear pre-conditions, which must be fulfilled, if the corresponding block shall be executed. A pre-condition can be e.g. a specific received sign sequence. Within a block, sign sequences can be sent out via the communication medium or received.

### *4.3.2 Data Tree Editor*

To allow graphically interactive programming of information processing procedures within the device driver, the creation of data structures is necessary. With the help of the Data Tree Editor, data structures can be created, dialogue aided and saved in project files. These data structures can be displayed in a graphic data structure overview and overtaken with a simple mouse click in the necessary processing dialogues.

### *4.3.3 Data Block Editor*

With the Data Block Editor functions, can be programmed to copy data in converted formats from a data stream to device driver specific data structures and vice versa. These functions are mainly used for developing encoding and decoding procedures for packets.

### *4.3.4 Key editor*

The task of the Key Editor is to assign clear recognition keys to specific packets so that an automatic identification for the decoding of the packets and further processing becames possible.

### *4.3.5 MMS Object Editor*

The MMS object editor allows one to create MMS communication objects, which handle the object related information processing within the device driver. For the modeling of internal information processing, transaction nets are used. All the MMS communication objects own one transaction net. The task of the transaction net is to request necessary information from the attached device for further processing in single actions with the help of the device-specific communication medium.

There are differences in the identification of the various MMS object classes. MMS variables created with the editor have an unambiguous identifier and belong to specific local variables in the global data of the device driver. The attached transaction nets of variable objects are specifically for each object. This is different for the MMS object classes Domain and Program Invocation. Objects of these two MMS classes are created with different identifiers during run time and not during the programming of the device driver. Therefore the related transaction nets are common for each class and not for each instance.

### *4.3.6 Transaction Net Editor*

Protocol sequences can be programmed graphically and interactively with the Transaction Net Editor. These transaction nets are assigned to MMS communication classes or to single MMS communication objects. From examinations carried out at the WZL, elementary modeling elements were derived which can be used for the programming. If there is a special modeling element needed, which is not present in the editor, it can be designed with the help of user dialogues.

Concluding the MMS-3D Kit is a migration aid that allows one to program graphically and interactively MMS-based device drivers needed for the integration of manufacturing devices with vendor-specific communication interfaces in OSI-based communication networks. In the future this kind of integration work will no longer be needed and heterogeneity will no longer be a problem if all manufacturing devices are OSI-compliant.

## 5 CONCLUSION

Up to now there is a significant lack of case tools which are designed and applicable for FMS software. The systems presented in this paper support new approaches to the development of such case tools. CellDesign facilitates the development of customer specific cell control software, CASCADE provides a framework for shop floor control applications and the MMS-

3D KIT enables a user to integrate machine tools incompatible with the Manufacturing Message Specification. The tools presented help to reduce development time and cost as well as the implementation effort of individual software solutions. Future research activities will focus on increasing reusability of software and extending the functionality of the case tools and the provided class libraries.

## 6 REFERENCES

Booch, G. (1991) Object Oriented Design With Applications. The Benjamin/Cummings Publishing Company, Inc.

Chen, J.B. et.al. (1995) Pursuing safe polymorphism on OOP. in Journal of Object-Oriented Programming, March-April.

ESPRIT Consortium CCE CNMA (1995), CCE: An Integration Platform for Distriubuted Manufacturing Applications. Springer Verlag

Frey, V. (1992) Planung der Leittechnik für flexible Fertigungsanlagen. Dissertation Universität Karlsruhe.

Friedrich, A. (1992) Offenes DNC-kommunikationssystem für numerische gesteuerte Arbeitsmaschinen.Reihe 20 VDI Verlag

Gamma, E. et.al. (1995) Design Patterns - Elements of Reusable Object-Oriented Software. Addison-Wesley.

Kohen, E. (1986) Adaptierbare Steuerungssoftware für Flexible Fertigungssysteme. Dissertation, Aachen University of Technology.

Lange, N. (1993) Dezentrale universelle Steuerungsarchitektur für Flexible Fertigungssysteme. Dissertation, Aachen University of Technology.

Meyer, B. (1994) Reusable-Software - The base objectoriented component libraries. Prentice-Hall.

Pree, W. (1995) Design Patterns for Object-Oriented Software Development. Addison-Wesley ACM-Press.

Pritschow, G. (1991) Leit- und Steuerungstechnik in flexiblen Produktionsanlagen. Carl-Hanser Verlag, München.

Schmid, H.A. (1994) Kundenspezifische Software zur Fertigungsautomatisierung durch Wiederverwendung eines objekt-orientierten Baukastens. CIM Management 6/94 pp.50-54

Rumbaugh, J. et.al. (1991) Object-Oriented Modelling and Design. Prentice Hall.

Stroustrup, B. (1992) The C++ programming language. Addison-Wesley.

Weck, M. (1995) Werkzeugmaschinen-Fertigungssysteme Band 3.2, Automatisierung und Stuerungstechnik. VDI-Verlag, Düsseldorf

Wirfs-Brock, R. et.al. (1990) Designing Object-Orinted Software. Prentice Hall.

Zipper, B. (1994) Das integrierte Betriebsmittelwesen - Baustein einer flexiblen Fertigung. Forschungsberichte iwb, Springer-Verlag, Berlin.

## 7 BIOGRAPHY

The Authors:
Prof. Dr.-Ing. Dr.-Ing. E. h. Manfred Weck, born in 1937; head of the Chair of Machine Tools since 1973 and member of the directorate of the laboratory for Machine Tools and Production Engineering (WZL) of the Rheinisch Westfälische Technische Hochschule Aachen (RWTH).
Dipl.-Ing. Jörg Friedrich, born 1964, studied Mechanical Engineering at the RWTH Aachen, engaged in scientific research at the WZL since 1991, Chair of Machine Tools, Controls of Manufacturing Systems group.
Dipl.-Ing. Thomas Koch, born 1963, studied Mechanical Engineering at the RWTH Aachen, engaged in scientific research at the WZL since 1989, Chair of Machine Tools, Controls of Manufacturing Systems group.
Dipl.-Ing. René Langen, born 1968 studied Mechanical Engineering at the RWTH Aachen and the Center for Advanced Manufacturing, Clemson S.C., USA, engaged in scientific research at the WZL since 1994, Chair of Machine Tools, Controls of Manufacturing Systems group.

# 11

# An Environment and Algorithm for FMS Controller Testing[†]

*Z. Deng*
*Narvik Institute of Technology*
*Teknologiveien 10, 8501 Narvik, Norway*
*Tel: 47-769-22181     Fax: 47-769-44866*
*E-mail: Ziqiong.Deng@hin.no*

*Z. Bi and Y. Zhu*
*Division #503*
*Nanjing University of Science and Technology*
*Nanjing 210094, P. R. China*
*Tel: 86-25-4315615     Fax: 86-25-4431622*

**Abstract**

The authors have developed a physical environment for testing controllers of Flexible Manufacturing Systems (FMS). In this paper, firstly, the testing mechanism is discussed. Secondly, the architecture of the testing environment is illustrated. Finally, an algorithm based on a model built for testing the correctness of the series of control commands from a tested FMS controller is given. A modelling methodology called Structured Macro Petri Net (SMPN) conceived by authors is described briefly.

**Keywords**

Software engineering, flexible manufacturing system (FMS), control engineering, software testing, system modelling, Petri net (PN), structured macro Petri net (SMPN)

---

[†] This work is supported with the pre-research foundation by Science and Industry Committee of China.

# 1 INTRODUCTION

For many manufacturers of consumer goods, flexible manufacturing and assembly is the only way in which they can efficiently compete in the market-place with a range of product variants. This is because clients are increasingly looking for products tailored to their own needs rather than mass-produced products (Rembold, 1993). Therefore, for producing such product variants, flexible manufacturing systems (FMS) are required.

Flexible manufacturing systems typically consist of (Deng, 1989):

- several manufacturing *equipment* (machines) such as CNC machining centres, CNC measuring machines, washing machines, etc.;
- part transport and handling equipment such as automatic guided vehicle (AGV) together with part loading and unloading stations, central part buffers, and local part buffers dedicated to an individual manufacturing machine to carry out efficient *part-flow* (or called *job-flow*) tasks within the FMS system;
- tool transport and exchange equipment such as movable robots together with tool loading and unloading stations, central tool base, and local tool magazines for each manufacturing machine to carry out efficient *tool-flow* within the FMS system.

All actions that happen among the part-flow procedure or the tool-flow procedure in an FMS are controlled by the *FMS controller*. In other words, the FMS controller makes a series of decisions and issues a series of commands to control the part-flow and/or the tool-flow in a series of time moments.

Since the early 1980s, there have been more and more FMSs installed in enterprises, and a number of FMS controllers developed by various developers. It seems that there exists a growing tendency of installing FMSs and developing various controllers for those FMSs.

As is well known, developing an FMS controller involves a complicated task in hardware and software development. Especially in the software development, if many bugs exist in the package, probably it may cause a disastrous result while the FMS is running. Therefore, the authors have launched a project since 1991 to develop a testing environment for testing the FMS controllers which may be developed by various developers. In other words, the authors wanted to create a testing centre where any developer or client of an FMS controller can ask the testing centre for the testing and/or debugging of his/her FMS controller. Meanwhile, if someone is willing to develop an FMS controller before his/her physical FMS system installation, he/she may also make use of the facilities of this testing centre to develop his/her FMS controller. It also means that we have to create a virtual equipment environment where the FMS configuration composed of CNC machines, part-flow sub-system, tool-flow sub-system, etc., can be configured for following purposes (Deng, 1995):

- receiving the sequence of control commands from an FMS controller being tested;
- executing the control commands;
- replying with a normal message to inform the FMS controller that a control command is executed properly or;
- replying with a abnormal message in random mode to inform the FMS controller that certain malfunction is happening in the facilities.

After years' work, the first stage of the testing environment is accomplished. In what follows in this paper, we will describe: (1) the mechanism of FMS controller testing, (2) the architecture of the testing environment, (3) the model and algorithm for testing real-time control function of a tested FMS controller, and (4) macro structurisation of places and resources.

## 2 MECHANISM OF FMS CONTROLLER TESTING

Generally, software production broadly follows the phases: *requirements, specifications, design, implementation, integration, maintenance, and finally, retirement* (Schach, 1990). After finishing the development of each of the first three phases, verification is required. After each of both the implementation phase and the integration phase, testing is required to assure the quality of that phase's solution. During the implementation phase's testing, the modules are tested. During the integration phase's testing, there are three types of testing required, namely *integration testing, product testing, and acceptance testing* (Schach, 1990). The purpose of integration testing is to check that the modules are combined together correctly to achieve a software product that satisfies its specifications. When the integration testing has been completed, product testing is performed. *The functionality of the product as a whole is checked against the specifications*. The final aspect is acceptance testing. Here the client enter the picture. The software is delivered to the client, who tests the software.

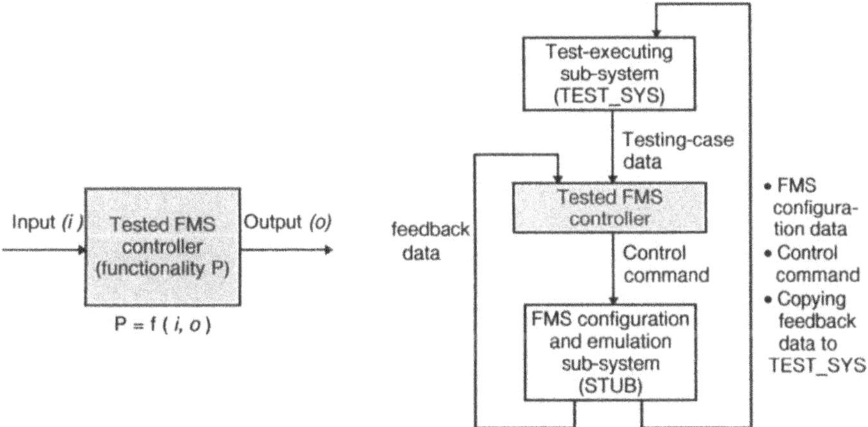


**Figure 1** Black-box testing for an FMS controller.

**Figure 2** Testing principle of FMS controller's testing.

As mentioned above in the introduction, our desire is to create a testing environment where developers of various FMS controllers can ask the testing environment for the testing of his/her FMS controller. Because a wide variety of FMS controllers are developed outside the testing environment, and the developments are carried out by various groups other than the group working in the testing environment, it is not usually possible for the group who works in testing environment to understand the internal software structure of the various tested FMS controllers. That is to say, our work is only *involved in product testing and/or acceptance testing* if a client wants to use the environment for his/her acceptance testing. Therefore, the mechanism conceived for FMS controllers' testing in our environment is limitedly to *the testing of functionality* which is specified in the specifications. In other words, the testing environment copes with a tested FMS controller as a black-box as shown in Figure 1.

Considering the relationship between the tested FMS controller and the testing system in our environment, the testing principle can be expressed and depicted as shown in Figure 2 where the testing system is composed of:

- a test-executing sub-system (or called TEST_SYS for brevity);
- an FMS configuration and emulation sub-system (or called STUB for brevity).

Initially, one should make use of the STUB (see lower part of Figure 2) to configure the FMS facilities, part-flow system, and tool-flow system which will be controlled by the tested FMS controller. The FMS configuration data created is then transferred to the TEST_SYS where the FMS configuration data is stored in the database and is ready for use by the TEST_SYS. According to both the functional specifications of the tested FMS controller and the stored FMS configuration data, the TEST_SYS creates various testing-case data (see upper part of Figure 2) to drive the tested FMS controller.

The content of the testing-case data includes the production data or, say, the job assignment which consists of the sets of data included in Tables 1 through 4.

**Table 1** Job assignment

| No. of job | Priority of job | Due date | Type of part | No. of blank | Batch quantity | No. of process plan |
|---|---|---|---|---|---|---|
| | | | | | | |

**Table 2** Process plan

| No. of process plan | No. of operation | No. of equipment | Alternative equipment | Priority | Sequence |
|---|---|---|---|---|---|
| | | | | | |

**Table 3** Operations

| No. of operation | No. of NC program | Duration of operation |
|---|---|---|
| | | |

**Table 4** Tool requirement

| No. of NC program | No. of tool | Type of tool | Tool life |
|---|---|---|---|
| | | | |

The FMS controller being tested starts to issue its first control command to the STUB according to the testing-case data. Meanwhile, the STUB emulates the execution of the control command, copies the control command to the TEST_SYS, transmits to the tested FMS controller either a normal feedback data, or an abnormal feedback data, and copies as well the feedback data to the TEST_SYS. From the second command on, the FMS controller issues its control commands according to not only the testing-case data, but also the feedback data.

# 3 ARCHITECTURE OF TESTING ENVIRONMENT

Creation of a testing environment involves architectural consideration concerning both the hardware environment and the software environment. The architectural consideration for the hardware environment should address the requirement for serving various sources of FMS controllers which may reside on various types of computers such as Sun workstations, VAX workstations, SGI workstations, HP workstations, and PC computers. Therefore, we have used the hardware configuration as shown in Figure 3.

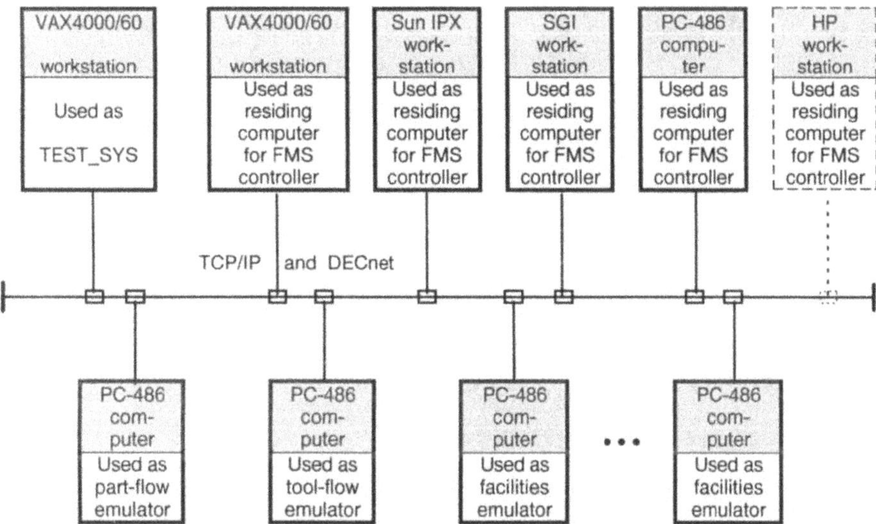


**Figure 3** Hardware configuration of testing environment for FMS controllers

The testing-executing sub-system (TEST_SYS) shown in Figure 2 had been installed on one of the VAX 4000/60 workstations as shown in the upper left corner of Figure 3. The STUB had been installed on several PC-486 computers as shown in the lower part of Figure 3. To serve as the residing computers for various sources of FMS controllers, one Sun IPX workstation, one SGI workstation, one PC-486 computer, together with one other VAX 4000/60 workstation had been installed. One HP workstation is planned to be installed soon.

The architectural considerations for the software environment had led us to adopt the NAS (Network Application Support) platform from Digital Equipment Co. as shown in Figures 4 and 5 (Digital Equipment Co., 1990).

The NAS platform can be used to support a compatible network of multivendor products. As shown in Figure 4, it provides compatibility among different operating systems such as UNIX, VMS, OS/2, MS-DOS, and etc.. As well, it provides compatibility among different databases such as RDB, Ingres, SyBASE, and Oracle. Regarding NAS, we find that it is suitable for meeting our objectives for testing various types of FMS controllers which may be required to work on different computers and databases.

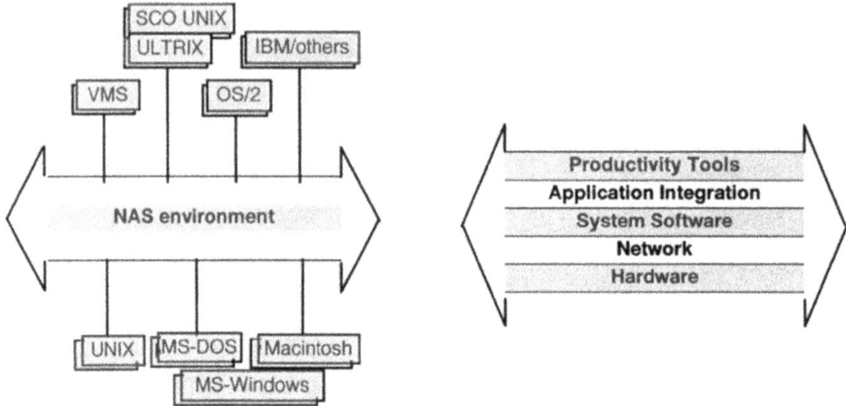


**Figure 4** NAS environment for multivendor support

**Figure 5** NAS environment for layers

# 4 ALGORITHM FOR TESTING REAL-TIME CONTROL FUNCTION OF A TESTED FMS CONTROLLER

As shown in Figure 2, the TEST_SYS has to

- create a complete set of testing-case data;
- check the correctness of the series of control commands both in normal and in abnormal situations;

and the STUB has to

- emulate the FMS operation;
- generate either normal feedback data, or abnormal feedback data.

Obviously, we have to conceive algorithms for realising each of the above tasks. In this paper, we only introduce an algorithm for correctness checking of control commands. Other algorithms can be referred to Huang (1995).

To check the correctness of the series of control commands, at first there is a need to create an analytical behaviour model for the TEST_SYS. This model should

- be independent of models which work in tested FMS controllers;
- exactly express the dynamic behaviour of FMS systems which are controlled by tested FMS controllers;
- be as simple as possible for easing the modelling and/or the model extension when needed.

We had conceived a modelling methodology for the TEST_SYS based on Petri-net (PN) theory and methodology (Peterson, 1981). As is well known, PN methodology makes use of *places* and *transitions* to graphically build a model in which tokens flow. If we consider the TEST_SYS model where attributes of tokens are related to concrete resources such as jobs (or say, parts), tools, manufacturing machines, etc., as one probably does normally, then here we may encounter the difficulty of unacceptable scale and complexity of the model built with the quantity of resources increasing.

To avoid such problems, we find that there exists a limited number of basic event-types which may happen in an FMS system as follows:

- in job-flow: a job is loading to or unloading from the FMS;
  a pallet together with job is transported from one machine to another machine;
  a pallet together with job is transported from a buffer to a workstation;
  a pallet together with job is transported from a workstation to a buffer;
  a machining operation is started or finished;
- in tool-flow: a tool is loading to or unloading from the FMS;
  a tool is transported between tool loading/unloading station and a local tool magazine;
  a tool is transported between tool loading/unloading station and central tool magazine;
  a tool is transported between a local tool magazine and central tool magazine;
  a tool is transported between two of local tool magazines.

Therefore, we firstly structure the places for the TEST_SYS's model, then define the structure of transitions. That is to say, we firstly defined a modelling methodology called SMPN (structured macro Petri-net) methodology with a procedure as follows:

- Resources are classified according to features of their dynamic attributes;
- The sets of places and transitions are structured abstractly to reflect the basic events and related system statuses;
- The global structure of the whole model is then built based on the cause-effect relationship of those basic events.

Then, using the above modelling methodology, a SMPN model for our testing system was built. In this paper, we only introduce the sub-model for job-flow as shown in Figure 6. The meanings of various acronyms of both places and transitions in the Figure are explained as follows:

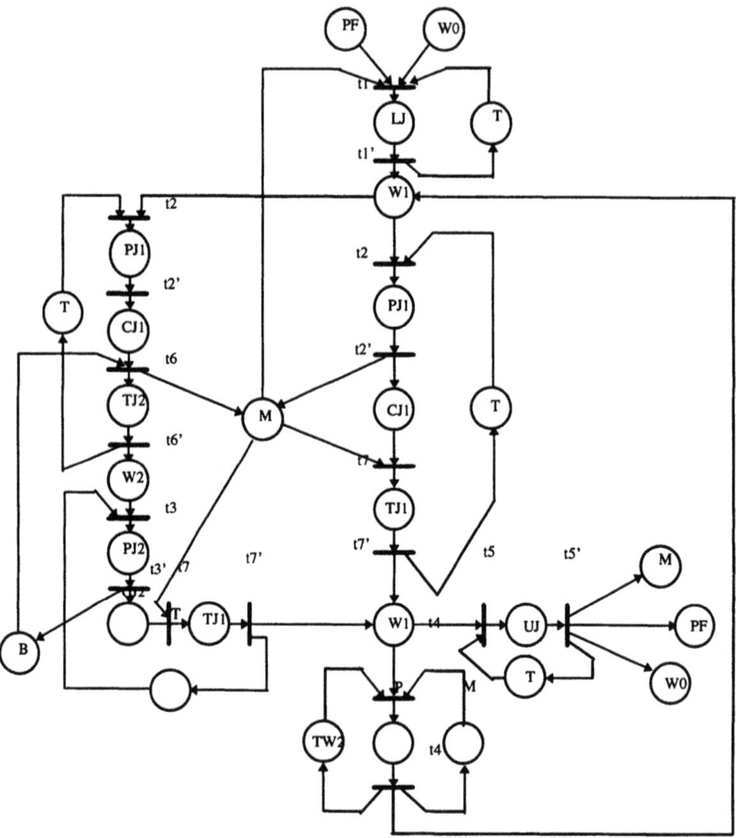


**Figure 6** Structured macro Petri-net (SMPN) model for job-flow of FMS for TEST_SYS.

- for *transitions*,

| | |
|---|---|
| t1, t1' | starting event (t1) and finishing event (t1') of a job-loading process; |
| t2, t2' | starting event and finishing event of a process of picking a pallet from a machine; |
| t3, t3' | starting event and finishing event of a process of picking a pellet from a buffer; |
| t4, t4' | starting event and finishing event of a job-machining process; |
| t5, t5' | starting event and finishing event of a job-unloading process; |
| t6, t6' | starting event and finishing event of a process of sending a pallet to a buffer; |
| t7, t7' | starting event and finishing event of a process of sending a pallet to a machine. |

- for *places*,

| | |
|---|---|
| LJ, UJ | processes of loading (LJ) and unloading (UJ) a job; |
| PJ1, PJ2 | processes of that a conveyer (AGV) picks up a pallet from a machine (PJ1) or from a buffer (PJ2); |
| TJ1, TJ2 | processes of that a conveyer (AGV) sends a pallet to a machine (TJ1) or to a buffer (TJ2); |
| P | process of a job-machining; |
| W0 | status of a job which is outside the system; |
| W1, W2 | statuses of a job which is on a machine (W1) or on a buffer (W2); |
| CJ1, CJ2 | statuses which represent that a conveyer has got a pallet from a machine (CJ1) or from a buffer (CJ2); |
| TW2 | status of a tool in a local magazine; |
| PF, T, B, M | available statuses of a fixture (PF), a tool (T), a buffer (B), and a machine (M). |

For brevity, we do not introduce the tool-flow model in this paper. It can be referred to Bi (1995) for details.

# 5 MACRO STRUCTURISATION OF PLACES AND RESOURCES

## 5.1 Methodology

- First, structure resources
- then, structure places

## 5.2 Structures of Resources describe the statuses of resources.

- **S1:** Pallet resource

{

| | |
|---|---|
| Code of pallet-type | * Different job may be assigned different pallet-type. |
| Flag of available status | * available, unavailable, or malfunctioned |
| Code of position where the pallet may reside in | * Code '0' means that the pallet resides outside the system |
| Code of pallet status | * occupied or not occupied |
| Job-code on pallet | * when the pallet is occupied by the job |
| Start-time of present status | |

}

- **S2:** Position Resource which can be used for Pallet Residing in (PRPR)

{

| | |
|---|---|
| Type-code of equipment which the position is belonged to | * including types of loading/unloading (L/U), buffer, machining, and transportation equipment. |
| Flag of available status | * available, unavailable, or malfunctioned |
| Code of equipment which the position is belonged to | |
| Code of position status | * occupied or not occupied |
| Code of pallet when it occupies the position | |
| Start-time of present status | |

}

- **S3:** Tool resource which includes L/U tool, transportation tool, and machining tool

- **S31: L/U tool resource**

| { | |
|---|---|
| Code of tool-type | * Different equipment may be assigned different type of U/L tool. |
| Flag of available status | * available, unavailable, or malfunctioned |
| Code of tool status | * in use or not in use |
| Code of equipment which is using the tool | |
| Start-time of present status | |
| } | |

- **S32: transportation tool resource**

| { | |
|---|---|
| Code of tool-type | * Different pallet may require different type of transportation tool. |
| Flag of available status | * available, unavailable, or malfunctioned |
| Code of tool status | * in use or not in use |
| Code of pallet which is using the tool | |
| Start-time of present status | |
| } | |

- **S33: machining tool resource**

| { | |
|---|---|
| Code of tool-type | * Different job may require different type of machining tool. |
| Flag of available status | * available, unavailable, or malfunctioned |
| Code of tool status | * in use or not in use |
| Code of equipment which is using the tool | |
| Start-time of present status | |
| } | |

- **S4: Job resource**

| { | |
|---|---|
| Code of job-type | * It can be used for searching the information of job machining process. |
| Code of finished machining operation within the machining process | |
| Code of pallet which is used by the job | |
| Start-time of present status | |
| } | |

- **S5: Workstation resource**

| { | |
|---|---|
| Code of workstation-type | * including types of loading/unloading (L/U), machining, washing, and measuring workstations. |
| Code of L/U tool-type required by the workstation | * special code when not required |
| Code set of PRPR which the workstation can supply | * It means how many PRPR the workstation possesses. |
| Flag of requiring machining tool | * required or not required |
| Occupying-status of machining position | * A machining position, work-table of a machine for example, is occupied by a pallet or not. |
| Code of job which is occupying the machining position | |
| Start-time of present status | |
| } | |

## 5.3 Structures of Places can be derived from structures of resources.

- **Structure PF: (S1 → PF)**

| { | |
|---|---|
| Code of pallet-type | * Different job may be assigned different pallet-type. |
| Code of L/U position where the pallet resides in | * Code '0' means that the pallet resides outside the system |
| Start-time of present status | |
| } | |

- **Structure of W0: (S4 → W0)**

{
Code of job-type
Start-time of present status
}

- **Structure of M: (S2 → M)**

{
Code of equipment which the position is belonged to
Start-time of present status
}

- **Structure of T: (S31 → T)**

```
{
Code of tool-type
Code of equipment which is using the tool
Start-time of present status
}
```

- **Structure of LJ: (S1, S31, S4, and S5 → LJ)**

```
{
Code of job-type
Code of L/U position
Code of L/U tool being used
Code of pallet-type
Start-time of L/U process
}
```

- **Structure of PJ1: (S2 and S32 → PJ1)**

```
{
Code of transportation tool-type
Code of pallet which is using the tool
Code of position which the pallet resides in
Start-time of present status
}
```

- **Structure of W1: (S2 → W1)**

```
{
Type-code of equipment
Flag of available status
Code of position status
Code of pallet when it occupies the position
Start-time of present status
}
```

- **Structure of CJ1: (S2 and S32 → CJ1)**

```
{
Code of transportation tool-type
Code of pallet which is using the tool
Code of position which the pallet resides in
Start-time of present status
}
```

## 5.4 Example

An FMS configuration shown in Figure 7 is composed of:

- a L/U workstation with manual L/U operation;
- a CNC machining workstation possessing two local positions for pallet residing;
- a washer workstation possessing without local position;
- a buffer consisting of four positions for pallet residing;
- an AGV with one position for pallet residing;
- three jobs a, b, and c waiting for entering the system;
- two pallets with one at L/U workstation and the other at one of the four positions of the buffer.

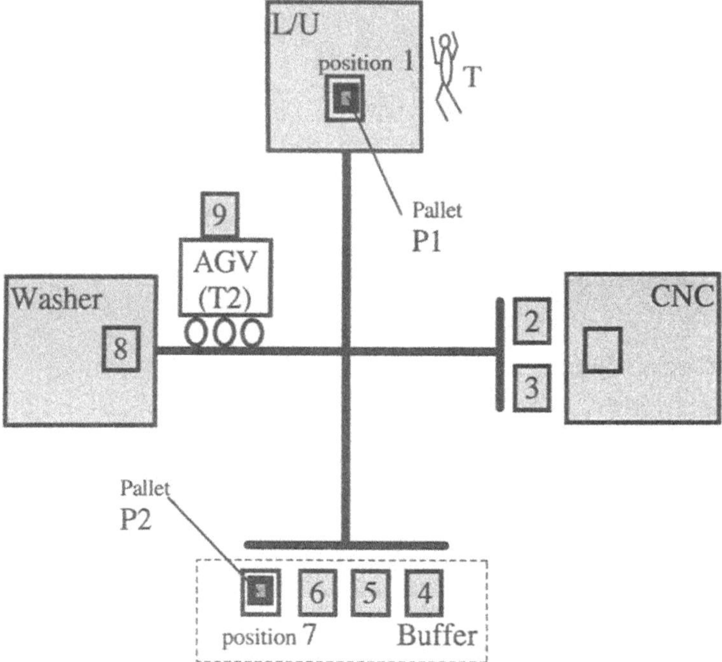


**Figure 7** An exemplified example of FMS

**After initialisation**

- Numbers of resource structures, S1, S2, S3 (S31, S32, S33), S4, S5 are

  2, 9, 2 (1, 1, 0), 3, 3

- Number of place structures (tokens), PF, W0, M, T, LJ, W1, PJ1, CJ1 derived from resource structures are

  1, 3, 1, (1, 1, 0), 0, 0, 0, 0

**t1 firing** (start to load job 'a')

- Job 'a' is loading.
- Token numbers in the input places of transition t1 - PF, M, and T are all decreased by 1 and become to be

  0, 0, (0, 1, 0).

- Token number in the output place LJ increases with 1 and is now equal to 1.
- The value of the token structure becomes to be

  LJ: {a, 1, T1, P1, t1}

**t1' firing** (finish the loading of job 'a')

- Token numbers in the input places of transition t1' - LJ is decreased by 1 and become to be 0 now.
- Token numbers in the output places of transition t1' - W1 and T both increase with 1.
- The values of the token structures become to be

  W1: {L/U workstation, available, occupied, P1, t1'}
  T: {T1, L/U workstation, t1'}

**t2 firing** (transportation tool T2 start to fetch pallet P1 with job 'a' loaded on it)

- Token numbers in the input places of transition t2 - both W1 and T are all decreased by 1 and become to be 0 and (1, 0, 0) now.
- Token number in the output place of transition t2 - PJ1 increase with 1 and becomes to be 1.
- The value of the token structure becomes to be

  PJ1: {T2, P1, 1, t2}

**t2' firing** (job 'a' has been fetched onto the transportation tool)

- Token number in the input place of transition t2' - PJ1 is decreased by 1 and become to be 0 now.
- Token number in the output place of transition t2' - CJ1 increase with 1.
- The value of the token structure becomes to be

CJ1: {T2, P1, 9, t2'}

## 6 CONCLUSION

The SMPN modelling methodology conceived by authors gains the benefits of: (1) the structured places and transitions represent a high level of abstraction; the values of structures of tokens are able to contain more information than that contained by tokens of conventional Petri net method; (2) the SMPN model relates only to the basic events that happen in an FMS, therefore there are no influences to the scale and complexity of the model when the quantity of jobs, devices, buffers, and alternative machining routes of jobs increases. Consequently a succinct SMPN model can be built.

The testing environment created by the authors is now running at its beginning stage. A few FMS controllers have been tested in this environment as case-studies and test-beds for the environment. It is found that the environment is capable of finding faults within the tested FMS controllers covering the following items: (1) unreasonable or false transition commands when simulating both normal situations and resources malfunctions, (2) violation of dispatching-priority rules, and (3) deadlock in job-flow and tool-flow.

For an FMS controller, one other functionality is the scheduling capability. The authors have not finished the research on how to test the scheduling capability yet. This is what the authors want to do in the next stage.

## 7 REFERENCES

Bi, Z. and Zhu, Y. and Deng, Z. (1995) Theory and Application of SMPN Modelling Methodology and Algorithm for FMS-controller Testing, Scientific report #20, Nanjing University of Science and Technology.

Deng, Z. and Wang, L. and Liu. X. (1989) A Study of Modelling Part Flow and Tool Flow for FMS, Software for Manufacturing, North-Holland.

Digital Equipment Co. (1990) Manufacturing Enterprise Handbook.

Huang, X. and Deng, Z. (1995) Algorithm for Generation of Testing-case in FMS-controller Testing, Scientific report #10, Nanjing University of Science and Technology.

Peterson, J. L. (1981) Petri Net Theory and the Modelling of Systems, Prentice-Hall, Inc..

Rembold, U. and Nnaji, B. O. and Storr, A. (1993) Computer Integrated Manufacturing and Engineering, Addison-Wesley.

Schach, S. R. (1990) Software Engineering, Aksen Associate, Inc..

Zhu, Y. and Bi, Z. and Deng, Z. (1995) Testing Mechanism for FMS-controller Testing, Scientific report #10, Nanjing University of Science and Technology.

12

# Reusability of Function-oriented and Object-oriented Master Control Software

*J. Uhl, J. Driller*
*Institute of Control Technology for Machine Tools and Manufacturing Units (ISW)*
*Seidenstr. 36, D-70174 Stuttgart, Germany;*
*Tel. +49 711/121-2420, Fax +49 711/121-2413;*
*Email: [joachim.uhl|juergen.driller]@isw.uni-stuttgart.de*

**Abstract**

Master control systems of the 5th generation are characterized by a change in paradigm in manufacturing technology. This change in paradigm is adapted automation, decentralized or human-centered. The necessary application flexibility and the possibility of forming variants require a high reusability of master control software, as well as a systematic development process. This paper shows as an example the software techniques which have been selected to develop a decentralized structured object-oriented master control system. A comparison is made with function-oriented master control systems.

**Keywords**

Object-orientation, master control system, software technology, reusability

## 1 INTRODUCTION

Control software must be viewed as a machine element which is to run through a systematic design and production process. Methods, procedures and development tools are used in this software design and production process (Siewert et al., 1994). This is generally called Software Engineering. However what is understood by software technology is application-oriented software engineering which uses methods and procedures tailored to the application.

In the development of control software, a differentiation is made between the application of software techniques for PLC and for NC control software in the machine control level and for master control software in the master control and cell control level. This paper is concerned with software techniques in the master / cell control level.

Apart from the general requirement of creating control software with systematic, engineering procedures and methods, the development of new software structures for master control systems of the 5th generation (Storr, Uhl, 1995; Pritschow et al., 1995; Driller, 1995; Adiga, 1993; Nof, 1994; Veeramani et al., 1993) influences in particular the development and application of software techniques for the master / cell control level.

## 2 REQUIREMENTS MADE ON MASTER AND CELL CONTROL SYSTEMS AND THE REQUIREMENTS RESULTING FROM THESE MADE ON THE SOFTWARE TECHNOLOGY

Requirements made on the 5th generation of master and cell control systems are an **individual automation adapted for each application**, a high flexibility, a simple adaptation to changes in in-house order control and resources (in particular manufacturing equipment) and a high reusability with the option of forming variants for a favorable cost-profit ratio of the master control software (Storr, Uhl, 1995; Uhl, 1995; Driller, 1995). These requirements can be fulfilled by (Uhl, 1995; Driller, 1995):

- a new software structured for control systems which is essentially reflected in engineering "physical" equipment and
- a development of master system applications based on a software modular system.

Figure 1 compares the software structure of a conventional master control system of the 4th generation (Brantner, 1993; Siewert, 1994) with function-oriented program modules and a database separated from the program modules of the decentralized, object-oriented software structure of a master control system of the 5th generation.

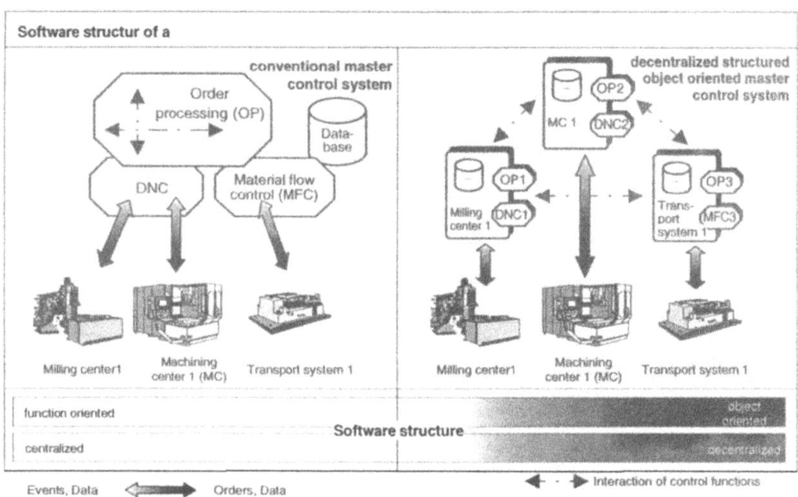


**Figure 1** Comparison of software structures

A **decentralized software structure** must be differentiated conceptually from a **decentralization in machine technology** and a **decentralization in an in-house order control (organizational)** and refers to the division of coordinating master control functions such as order execution of software objects. Apart from coordinating master control functions, a master control system also possesses executive master control functions such as the DNC master control function.

While mostly traditional procedures (e.g. application of the waterfall model) and methods (e.g. structured analysis, entity-relationship, structured design, etc.) of software engineering were used for master control systems of the 4th generation, then in the master control systems of the 5th generation, methods and procedures adapted correspondingly are required in order to create time-effectively an inexpensive software.

The methods and procedures must guarantee (Storr, Uhl, 1995; Uhl, 1995; Driller, 1995):

- a structuring of the software as per flexible production system (FPS) components with a systematic, engineering decomposition of the FPS components and the representation of these relationships (see also Fig. 2),
- a structuring of the software according to master control functions,
- high consistency of the description methods used, i.e. a consistent application of information in all phases of the software engineering process,
- support of a simple reusability of the master control software from a software modular system,
- master control software which is easy to understand.

Figure 2 shows an example of the structure of the DNC master control function for a manufacturing system with two machines. The objects of the representation have been derived from a physical view of the manufacturing system to be automated.

A library is used as development platform to support the development work. The platform contains software for **communication** (between objects and also between processors), a **database** and **help functions** (e.g. list administration). The library will be used by all software developers for developing master control software. A commercially supported platform must be used. The main advantage of such a platform is the guarantee of computer independence arising from the encapsulation of operating system routines etc. This relieves software developers of routine programming work and he/she can then concentrate on the actual application. Another advantage is that the library routines will have been well tested, so software quality should improve.

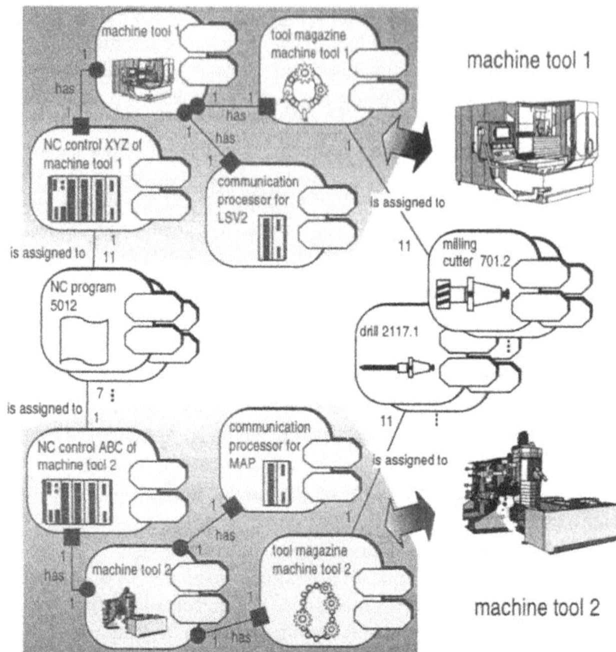


**Figure 2** Example of the structure of the DNC master control function

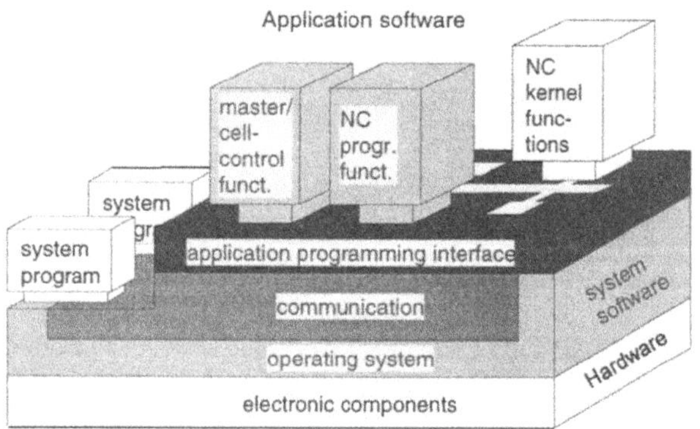


**Figure 3** Application and structure of a platform

# 3 SOFTWARE TECHNOLOGY FOR MASTER CONTROL SYSTEMS OF THE 5TH GENERATION

## 3.1 Procedures for creating a master control system based on a modular software system

Development of a master control software using systematic design is required. A master control system of the 5th generation structured according to objects offers a good starting point due to its orientation towards "physical" equipment. This orientation facilitates the "design" of future master control systems for specific applications. The basis for this approach is the application of procedures and methods of object-oriented software engineering.

Figure 4 shows a procedure for developing an application-specific master control system in two stages. The development of the **master control software** is to be differentiated from the generation of the **master control system**. The procedure can be classified in the object-oriented software life cycle with the phases system analysis, system design, coding and module test as well as system test and use. The aim is the development of classes of the **class library of the master control software**. The classes of the **class library of the development platform** which are usually only available as source or binary code must already be taken into consideration in the module design and are included using reverse engineering.

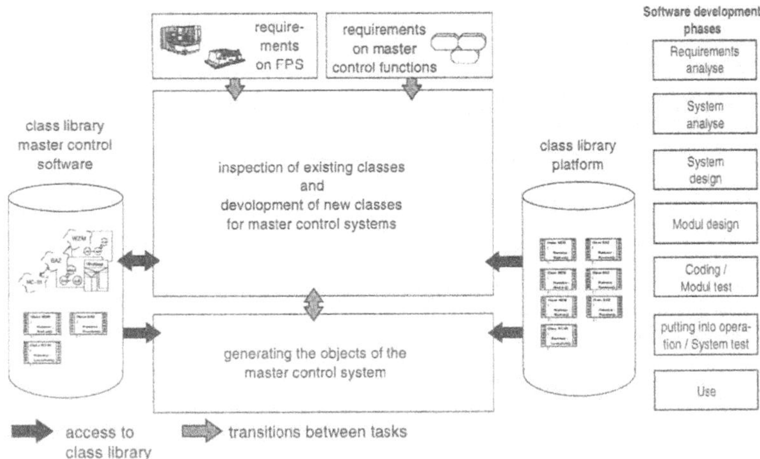


**Figure 4** Procedure for creating a master control system

Figure 5 shows how the phases of the software engineering process will proceed. The **initial development** and the **follow-on development** of a master control system are differentiated as well as an **incomplete** and a **complete master control system class library** for generating the objects of a master control system.

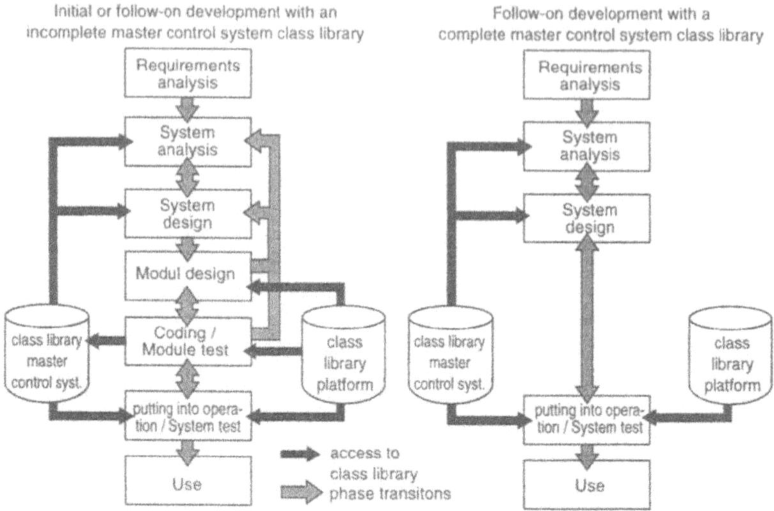


**Figure 5** Differentiation of a development with an incomplete and a complete class library for the classes of the master control software

If the class library does not exist for the classes of the master control system in an initial development or is not complete in a follow-on development for a different application, all phases of the software engineering must be executed. A new development or adaptation of classes of the master system class library is necessary. If the control system can be created completely from the master system class library for a follow-on development for another application, then phases of the software engineering process as seen on the right of Fig. 5 can be skipped. Therefore in the system design phase, for example, the objects required and instantiated from the class library for the master control system only have to be checked and defined.

Figure 6 describes the systematic steps for developing a master control system. Note that the application-specific steps

- requirements made on FPS components and
- requirements made on master control functions

are the starting point. The transitions between the steps are progressive. The procedure includes frequent forward and backward movement between the steps.

The following are essential techniques in system analysis and system design:

- Decomposition of executive master control functions and FPS components
- Decentralization of coordinating master functions to FPS components.

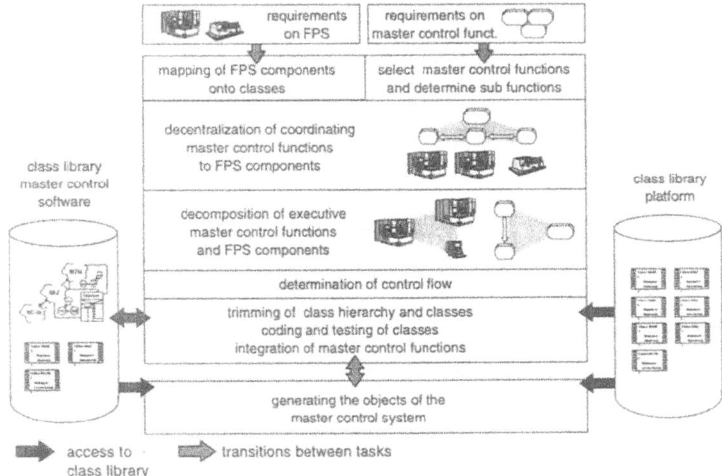


**Figure 6** Systematic steps for developing classes for a master control software

## 3.2 Descriptive methods used

The descriptive methods that are used have a considerable influence on the consistency achieved between a system analysis, system design and coding. The development platform has an influence on the consistency between module design and coding. An important prerequisite for consistency is that an object-oriented descriptive method and an object-oriented development platform are used, i.e. that objects, classes, inheritance, aggregation and association are supported.

The **descriptive method of G. Booch** (Booch, 1991) has been used in the core phases, where actual software generation occurs, due to:

- its high consistency up to the option of code generation,
- a suitable notation (e.g. possibility of representing objects, classes, inheritance, aggregation and association relationships)
- its intelligibility etc.

In the following discussion it is assumed that the reader is familiar with the Booch notation. This however has been extended by **Nassi-Shneiderman diagrams** for the structured formulation of the methods of a class. Figure 7 shows the descriptive methods used in the various phases.

Figure 8 shows an example of the application of the descriptive methods for the decomposition of the FPS components of the machining center together with the DNC master control function from Figure 3. The decomposition is made corresponding to the decomposition during the **development of PLC software** in the machine control level. State diagrams are assigned to each class. Dynamic behavior, however, is not implemented as the switching between status. but by method calls on the objects in the object diagram, such as those in Figure 10.

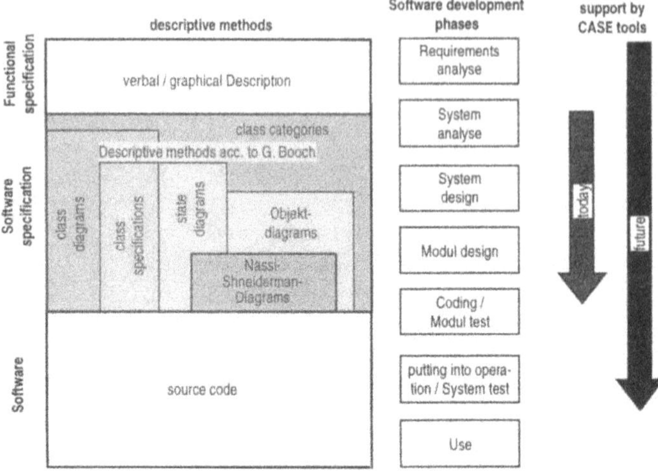


**Figure 7** Descriptive methods used

Figure 9 and Figure 10 show examples of the use of a software tool for the descriptive method according to Booch. In Figure 9, the master control functions of a master control system are shown in class categories. The classes of the DNC class category are shown in Figure 10. The representation of a run for transferring an NC program to a machining center can be see in Figure 11.

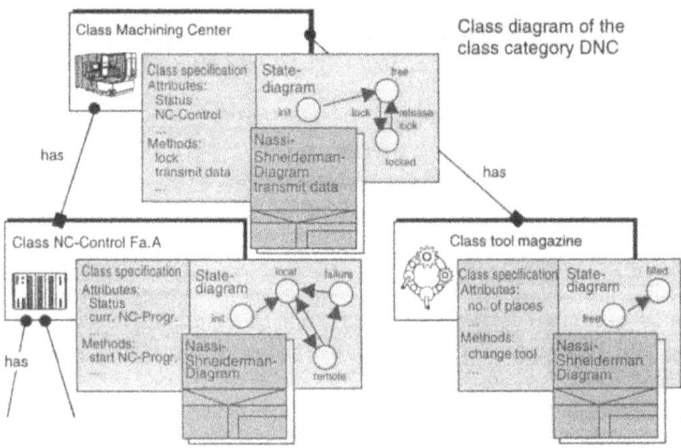


**Figure 8** Engineering decomposition and application of descriptive methods for representing a machining center

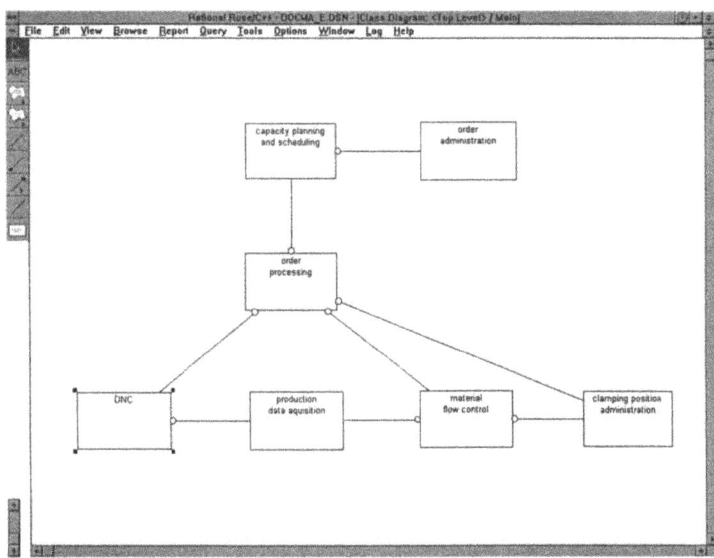


**Figure 9** Class categories in a master control system

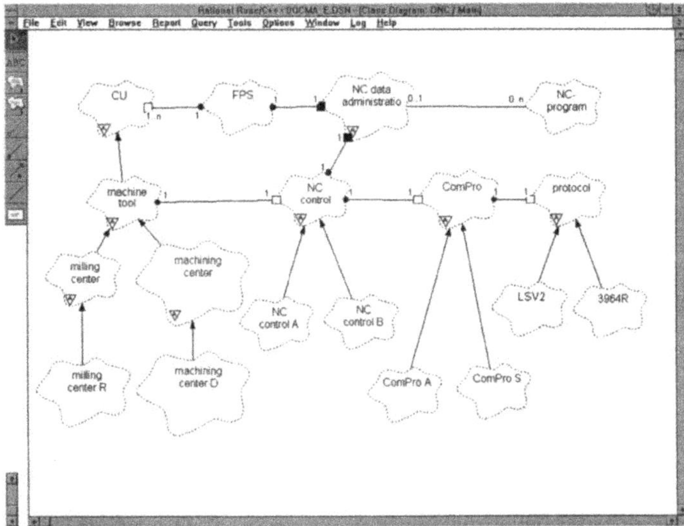


**Figure 10** Example of a class diagram for DNC master control function

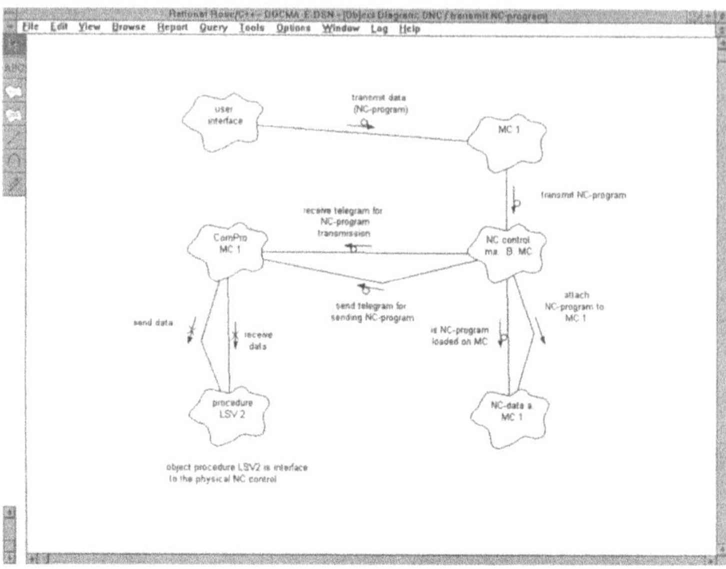


**Figure 11** Object diagram for the execution of the transfer of an NC program

A standardized platform such as a platform based on the **CORBA standard** (Common Object Request Broker Architecture) or the **OSACA platform** (Open System Architecture for Controls within Automation Systems) should ultimately be used as the development platform. A development platform based on the CORBA standard has the advantage that it supports an object-orientation with objects, classes, inheritance, aggregation and association relationships across computers. An OSACA platform enables interoperability with objects at the machine control level.

## 4 IMPLEMENTATION EXAMPLE AND COMPARISON WITH FUNCTIONAL MASTER CONTROL SYSTEMS

The application of the software techniques discussed in this paper took place at ISW in the framework of prototype work on the decentralized structured, object-oriented production master control system for manufacturing applications (DOCMA). The CASE tool ROSE developed by Rational was used for modeling the master control software. The DSOM (Distributed System Object Model) platform developed by IBM based on the CORBA standard was used as the development platform. Figure 12 shows an implementation example.

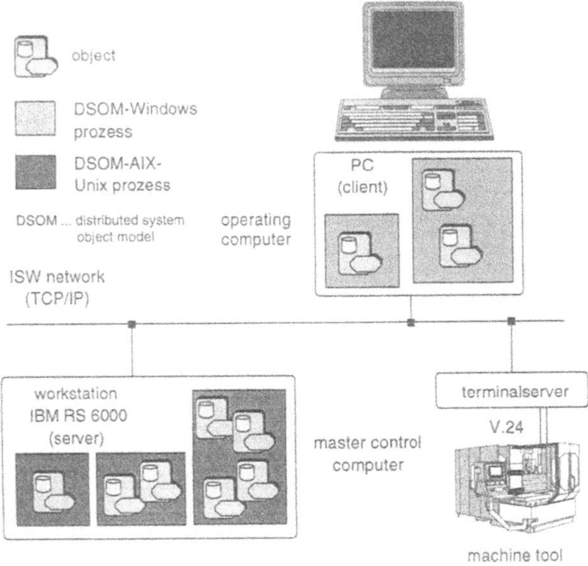


**Figure 12** Implementation example

Table 1 shows a comparison of the decentralized object-oriented master control system DOCMA with the adaptable master control system ALSYS. ALSYS is a master control system of the 4th generation and was also developed at the institute. The comparison is for the DNC master control function for an FPS with 2 different machines and 1 transport device. It addresses the software techniques used and the reusability of master control software.

| **Criteria** | **ALSYS** | **DOCMA** |
|---|---|---|
| Number of the phase transitions with a change of the descriptive method | Requirement analysis to system analysis, system analysis to system design, system design to coding | Requirement analysis to system analysis, system design to coding |
| Descriptive methods in system analysis:<br><br>System design:<br><br>Coding: | Structured Analysis, Entity Relationship<br>Structured Design, Structograms<br>C | Descriptive method acc. to G. Booch, Nassi-Shneiderman diagrams<br><br>C++ |

| | | |
|---|---|---|
| Reusability of master control software | Function modules | Classes, inheritance |
| Reusability of master control software by | Configuration data, parametrization, editable sequence rules, etc. | Classes, inheritance, parametrization |
| Extension of master control software by master control functions | New function modules, extension of function modules | New classes, inheritance |
| Extension of the master control software by FPS components | Extension of existing function modules | New classes, inheritance |
| Archiving | Versions of function modules various FPS applications | Class library |
| average number of function modules / classes \ objects of DNC | approx. 3 - 5 | approx. 14 - 18 \ 10 - 12 |
| average reusability of function modules / classes of DNC within new applications | approx. 0 - 5 | approx. 10 - 18 |
| average number of source code lines per function module / class | approx. 14 000 - 55 000 | approx. 100 - 2000 |

**Table 1** Comparison of software technology and reusability for master control systems of the 4th and 5th generation

The key observations from the use of object-oriented software techniques in DOCMA also compared to functional software techniques in ALSYS can be comprised in the following points:

- The master control system is developed using a consistent software model and descriptive form. The software module does not need to be transformed during the transition from system analysis to system design.
- The creation of a library with classes for master control system applications is achievable. Object-orientation (abstract data types, inheritance mechanisms, aggregation etc.) offers an appropriate mechanism for this. However the consistent use of the descriptive methods is necessary in order to produce detailed documentation. Aids for managing the increasing number of similar classes must also be provided (e.g. good search mechanisms).
- Software changes to facilitate reuse of function modules for new applications cannot be avoided. Despite a high adaptability in functional master control systems, function modules are not universal. An object-oriented master control system on the other hand achieves a

high reusability due to a high number of classes with lesser functionality which can be easily exchanged.

- The descriptive methods in all phases must be based on the same object-oriented model (same mechanisms, e.g. inheritance, etc.) otherwise a transformation of the software model is necessary. This relates in particular to the transition from module design to coding. Examples of such transformations arise when we use
  - C instead of C++ (inheritance mechanisms have to be reproduced exactly),
  - a platform with interprocess communication instead of a platform with distributed object-oriented communication (object communication has to be represented via interprocess communication) or
  - a relational database instead of an object-oriented database (data and methods have to be separated for persistent storage in the application).

## 5 SUMMARY

This paper has given an insight into the use of object-oriented software techniques and systematic design in the development of a 5th generation master control system. It has shown how a reusable class library can be created for decentralized object-oriented master control systems and how application-specific master control system can be generated from the class library. The advantages of "designing" an application-specific master control systems with an interdisciplinary, physical view of the master control system have become clear. A comparison with functional master control systems was made and a simpler and higher reusability of classes compared to function modules was observed.

For the development of decentralized, object-oriented master control systems, consistent support by CASE tools is not yet available. Development work is particularly needed in the setting-up phase. This concerns the development of test environments with whose help the production system to be controlled is reproduced and simulated in order to check the operability of a master control system.

## 6 REFERENCES

Adiga, S. (1993) (Ed.) Object oriented Software for Manufacturing Systems. London etc.: *Chapman & Hall*

Booch, G. (1991) Object-oriented Design with Applications. Redwood City: Benjamin/Cummings

Brantner, K. (1993) Adaptierbares Leitsteuerungssystem für flexible Produktionssysteme. *Dissertation Universität Stuttgart (ISW 96).* Berlin, Heidelberg, New York: Springer

Driller, J. (1995) Modellierung verteilter Steuerungs- und Informationssysteme in der Produktionstechnik. In: *Graduiertenkolleg Parallele und Verteilte Systeme, Zusammenfassender Bericht 1993-1994*, Stuttgart

Driller, J. (1995) Föderative Steuerungssysteme in der Produktionstechnik. *3. Berichtskolloquium des GK PVS*, Stuttgart

Nof, S.Y. (1994) Critiquing the potential of object orientation in manufacturing. *Int. Journal of Integrated Manufacturing*, Vol. 7, No. 1, 3-16

Pritschow, G.; Storr, A.; Handel, D.; Rommel, B.; Uhl, J. (1995) Uniform Object-Oriented Machining Model as Basis for Decentralized Planning in Master Control Systems. In: *Annals of the German Academic Society for Production Engineering*

Siewert, U.; Reichenbächer, J.; Uhl, J. (1994) Softwarewerkzeuge für die Steuerungstechnik. In: *Fertigungstechnisches Kolloquium (FTK) '94*. Berlin, Heidelberg, New York: Springer

Siewert, U. (1994) Systematische Erstellung adaptierbarer Leitsteuerungssoftware am Beispiel der Durchsetzungsplanung. *Dissertation Universität Stuttgart (ISW 100)*. Berlin, Heidelberg, New York: Springer

Storr, A.; Uhl, J. (1995) Objektorientierte Leittechnik: neue Perspektiven und Lösungen. *CIM-Management,* 11, 30 - 34.

Storr, A.; Reibetanz, T.; Uhl, J. (1995) Strukturierung von Zellenleit- und NC-Programmierfunktionen als Beitrag zu offenen Steuerungsystemen. In: *Offene Steuerungen,* München, Wien: Hanser

Uhl, J. (1995) Leittechnik von morgen - Was bringt die Zukunft? In*: Effiziente, prozeßnahe Fertigungssteuerung und Überwachung.* ISW-Eigenverlag

Uhl, J. (to be published) Beitrag zur Entwurfssystematik eines dezentral strukturierten, objektorientierten Leitsystems am Beispiel der DNC und Auftragsdurchsetzung. *Dissertation Universität Stuttgart*

Veeramani, D.; Bhargava, B.; Barash, M.M. (1993) Information system architecture for heterarchical control of large FMSs. *Computer Integrated Manufacturing Systems*, Vol. 6, No. 2, 76-92

## 7 BIOGRAPHY

Joachim Uhl holds a master's degree in mechanical engineering. Since 1990, he has been working as research assistant at the Institute of control technology (ISW) at the University of Stuttgart and is head of the group 'production control systems and quality assurance'. His main research interests are the application of object-oriented techniques in manufacturing, the development of decentralized control systems and the development of open system architectures for cell controls.

Jürgen Driller holds a master's degree in computer science. Since 1993, he has been working as a research assistant at the Institute of control technology (ISW) at the University of Stuttgart. He is also a member of the *Graduiertenkolleg* 'Parallel and Distributed systems'. His main research interests are the application of object-oriented techniques in manufacturing and the development of decentralized control systems.

# 13

# A Reusable Software Artifact Library system as the core of a reuse-oriented software enterprise

*G. Jacucci*, E. Mambella†, G. Succi*, C. Uhrik‡, M. Ronchetti*, A. Lo Surdo†, S. Doublait†, A. Valerio**

** Laboratorio di Ingegneria Informatica, DISA, Università di Trento, via F. Zeni 8, I-38068 Rovereto, Italia, tel. +39-464-443140, fax +39-464-443141*

*† Sodalia S.p.A., via Brennero 364, I-38100 Trento, Italia tel. +39-461-316111, fax +39-461-316663*

*‡ Faculty of the Dept. of Technology Programs, University of Phoenix 7800 E. Dorado Pl. , Englewood, CO*

*E-mail: gianni@lii.unitn.it*

**Abstract**

Introducing software reuse at a corporate level represents one of the most promising means of addressing the rising costs that are plaguing the software industry. A series of mechanisms are needed for shortening development cycles and providing reliable software of high quality which will be more maintainable and flexible for future extensions. This paper describes the experiences of Sodalia S.p.A., a young Italian company, in implementing such reuse methodology, particularly centred around a reuse tool specifically developed.

Since 1993, Sodalia's software engineers have been implementing a reuse program whose goal is making software reuse a significant and systematic part of the software process. The Sodalia Corporate Reuse Program is intended to institutionalize a software reuse process that incorporates reuse-specific activities all along the Sodalia object oriented software development process, drawing heavily on a reusable software artifact library system which has been designed to support the classification, management and search for artifacts to be employed in reuse efforts.

This paper presents an overview of the corporate reuse program implemented at Sodalia, focusing on the reusable software artifact library system and its role inside the reuse program.

**Keywords**

Software reuse, reusable software artifact library, reuse support organization

## 1 INTRODUCTION

Sodalia S.p.A. is a young Italian company that arose from a joint-venture between the STET Group (Italy) and Bell Atlantic Corporation (USA). The objective of Sodalia is the development of innovative telecommunications software products for the management and maintenance of telecommunication networks. The company's goal is to develop high quality, low-cost software and, more generally, methodologies which are able to deal with the even more complex and broad needs of the telecommunication sector, exploiting innovative development methods and technologies.

Due to the increasing competition and rapid technological innovations that in the past few years mark the field of telecommunications, this sector has undergone radical changes, forcing operators to cope with a growing demand of new applications and services, in terms of quality, variety, reliability, and, last but not least, low prices.

In this situation the traditional custom software development methodologies show all its inadequateness and ineffectiveness: to become really competitive in the telecommunication sector and able to survive in a global market, an organization needs to adopt leading edge technologies inside a defined, well planned and specific software development process. Moreover, traditional software development methodologies do not allow an efficient maintenance of the products, while any modification or extension of these products often results in a full re-development of them.

A new iterative development philosophy based on software reuse and using the emerging object oriented technologies seems to be the right trade-off among the opposing factors characterizing a successful telecommunication product or service: high-quality, low-cost, reduced time-to-market, flexibility and maintainability, just to name some. Software reuse is particularly interesting for Sodalia, since it develops similar applications for its parent companies and a new software system can be built reusing previously developed components.

Since 1993, Sodalia's Software Engineers have been studying a systematic reuse program to support and to integrate the software development activities, incorporating a reuse library to support the classification and management of the heterogeneous reusable components.

The next section of this paper deals with Sodalia's Corporate Reuse Program, while section 3 describes in detail the Reusable Software Artifact Library, highlighting its architecture and main functionality. Section 4 presents some experiences gained in the use of this reuse methodology inside Sodalia. In the last section some conclusions are outlined and future work is sketched.

## 2 SODALIA'S CORPORATE REUSE PROGRAM

The Sodalia reuse program aims at making software reuse a significant and systematic part of its iterative software development process. Due to the critical role played by reuse in achieving Sodalia's business objectives, and the novelty of the reuse technology, the reuse program will be monitored and updated as the reuse experience grows, iteration after iteration.

The reuse program aims to develop a Software Reuse Process that incorporates reuse-specific activities within the Object-Oriented Software Development Process, and a reuse library to support the classification and management of reusable components. The strategy adopted to support systematic reuse within and across iterations and projects, organizes activities around the following two views: **Software Process for Reuse** and **Software Process with Reuse**, supported by a **Reusable Software Artifact Library** (RSAL).

A Metrics Definition Program has been set up with the goal of defining metrics for both software and reuse processes (SSPG 1994b). The first objective of the program is the definition of size and complexity metrics for measuring source code quality. Then, the metrics model will be extended to cover the analysis and design phases. Finally, a predictive metrics model will be conceived. The following sections expose the details of the corporate reuse program.

### *The concept of reuse for Sodalia*

Beyond the apparent obviousness of the concepts and objectives of reuse, several competing reuse approaches differ in at least one aspect: the artifacts they intend to reuse (e.g., the reuse of software, or of all life-cycle work-products). The following definition has been adopted to capture the essence of the Sodalia reuse approach:

> Software Reuse is the set of planned and systematic activities aimed at maximizing the use of existing software artifacts and known processes in the production and maintenance of new software artifacts.

Thus, Sodalia's reuse activity considers every kind of artifact, although more importance is given to high-level artifacts of the software life-cycle (e.g., requirements), since they hold the maximum information (see (Kain 1994), (Capers 1994), (Prieto-Diaz 1993)).

### *A Reuse-centered Software Development Process*

Sodalia's Software Engineering departments have defined and applied a reuse-centered software development process that fully supports software reuse and exploits the benefits of object-oriented analysis, design, and programming.

SIMEP (Sodalia's Integrated Management and Engineering Process (SSPG 1994a)) provides an advanced process architecture, modelling a highly iterative process as well as rapid prototyping. The iterative process model provides a rigorous framework for progression in the project by re-iterating a sequence of basic process steps. Moreover, the process architecture identifies reuse-specific activities (e.g., Problem Domain Analysis, Generalization of Requirements, etc.) as well as reuse-affected activities (e.g., coding) to foster reuse at all stages of the production process. The reuse-specific activities are described in the Sodalia's Software Reuse Guidelines (SSRG), (SSPG 1994c).

The SSRG "supports" SIMEP (reuse in-the-process) to provide guidance for proper conduct of a set of planned and systematic activities carefully defined and positioned along the development process to:

- **Develop with reuse**: maximize the reuse of already existing software components and artifacts;
- **Develop for reuse**: produce software with the highest reuse potential.

The SSRG "complements" SIMEP (reuse in-the-factory) to provide guidance for all those activities, across and beyond projects, required to establish, maintain, and populate the RSAL, by producing specific components out of the projects. Moreover, the purpose of the guidelines for the activities of reuse in-the-factory, is to maximize the matching of reuse opportunity with reuse potential, through specific rules and recommendations to populate, search, extract, and maintain the reuse library.

### *Software process for reuse*

Software process for reuse includes the set of activities that allow, during the course of a project, the early identification of artifacts to be developed that might exhibit a high reuse potential either within or in other iterations of projects. Besides, added costs and time (incurred for reuse addressed specific activities) estimation techniques will be defined.

Sodalia pursues both **vertical reuse** (i.e., the reuse of software artifacts within a specific domain or application area) and **horizontal reuse** (i.e., the reuse of software artifacts across domains or application area). However, it puts major emphasis on the latter, since horizontal reuse provides the highest payoff.

Architecturally, Sodalia's products are viewed as a hierarchy of a number of product layers, each playing a well defined technical and market role. This architecture is defined by the Strategic Product Architecture (SPA) (SSPG 1993), whose goal is to set the strategic direction for structuring Sodalia's software products to support reuse activities. The SPA is organized in four layers:

- **Operating Environment** (layer 1): it addresses the technological foundation of the software products. The architectural components of this layer reflect the selection of basic database, graphics, and communications enabling technologies for the Operating Environment.
- **Application Shell** (layer 2): it characterizes software components specifically designed to solve application problems for a broad telecommunication market (e.g., Billing, Service Provisioning, Customer Network Management, etc.). This layer is likely to contain a large number of reusable software components.
- **Market Segment Specific** (layer 3): it reflects the specificity of a market segment related to a specific problem-domain (e.g., the Service Provisioning Application can be customized for small PTTs, large end-users, and mobile telephony market segments).
- **Customer Specific** (layer 4): it represents those software components which are highly specialized to address application features specific to a particular customer. Components at this layer have typically a low reuse potential.

The main architectural principle employed in SPA is relentlessly separating generic functions (i.e., that might exhibit a high reuse potential) from those which are specific to problem domains, market segments, and individual customers. The higher in the SPA a component is categorized, the less likely that component is to be reused. This concept is particularly true for horizontal reuse. However, within the same application domain (vertical reuse), components at all layers are equally highly reusable.

In application of the SPA, requirements for the components to be developed (at any level in the SPA) are examined to determine their reuse potential. The examination may recommend a "generalization" of a requirement to increase reuse potential. Since generalization should comply with market needs, it should be driven by stability and/or evolution analysis of the user requirements related to the market segment involved, as determined by the related business plan. This critical activity is driven by the Problem Domain Analysis. The lower the level of the component, the more general is the analysis. On the contrary, the higher the level of the component, the more Customer-Specific or Market Segment-oriented is the analysis.

### *Software process with reuse*

Software process with reuse includes the set of activities that aims at maximizing reuse of existing components classified into the RSAL.

The approach supports the earliest possible identification (at requirements or design stages) of candidate reusable components since reuse is most effective when applied at early stages of the life-cycle. The principle, well practiced by hardware engineers, is to specify and design aiming at reusing existing components. The earlier these components are identified, the greater is the possibility to tailor requirements and design to reuse those components, rather than developing new ones. Since the definition of reuse is based on applying existing solutions to new problems, one can succeed in identifying something to reuse only if:

- The domain of available solutions is complete with respect to new, emerging problems;
- The description of both the problem and the solution are expressed at similar levels of granularity, for assessing their match.

If either one of these basic conditions is violated, it is practically certain that a new solution will be developed even for an old problem. Thus, the issue of anticipating the identification of reusable solutions depends on how early in the life-cycle one can satisfy the above necessary conditions. As a result, the reusable asset library can only be populated by artifacts certified according to the above conditions.

The identification of candidate reusable components must take place at each iteration (SSPG 1994a), moving top-down from large-grain artifacts (e.g., sub-systems) to fine-grain artifacts (e.g., classes), thus supporting the "reuse in the large" and "reuse in the small" paradigms.

Reuse "as-is" (i.e., without any alteration) and reuse "with change" (i.e., requesting prior modification/generalization/composition) have also been considered. Our strategy focuses on maximizing reuse as-is rather than reuse with change, because it maximizes productivity, relies on a certified quality, and minimizes (by factorizing) maintenance costs. Nevertheless, reuse with change should be considered when it is the only viable alternative. The latter implies the reprocessing of intermediate life-cycle artifacts of the component (specifications, design, test, etc.) before being able to reuse it. The cost of developing reusable components can be amortized only if the components are used repetitively with no or minimal changes.

## 3 REUSABLE SOFTWARE ARTIFACT LIBRARY

RSAL is a system for organizing rationally **artifacts** of a heterogeneous environment targeted to software production. The RSAL system supports the classification and search of various artifacts associated with a family of software development projects. By artifact, we mean anything with a self standing identity that can be found in the software life cycle: reports, user expectations, requirements, cost models, algorithms, programs are non exhaustive examples of artifacts. An artifact can be defined as the final result of an activity, including documents (e.g., High Level Architecture, User Manual) as well as software products (e.g., sub-systems, classes).

A heterogeneous environment targeted to software production is by its own nature distributed in time and space, multi-systems and heterogeneous in terms of the machines and the operating systems working on it.

RSAL does not manipulate artifacts, but rather information on artifacts which are stored in software repositories, either local (i.e., owned by Sodalia) or remote (i.e., owned by Sodalia Parent Companies or others). Each artifact is described in RSAL by means of an artifact descriptor, whose two-fold purpose is to uniquely identify (locate) the artifact and to maintain a public description.

Artifacts have intrinsic attributes which pertain to their ontology, since they define their essence, e.g. the last opening date, the owner, the name, the i-node number are all intrinsic attributes of a Unix file. Artifacts have external attributes as well, which are annotations referring to artifacts and used in classifying and retrieving them: the quality level of a product, a description of the behaviour of a piece of code, a local comment on a remote file are all examples of annotations. RSAL handles both intrinsic attributes of its artifacts and external attributes. Moreover it allows to add further annotations to an artifact so as to make classification and search easier.

In RSAL there is a set of predefined attributes, i.e. attributes that are defined for any artifact. The set of such attributes includes the author, the creation date, the size and so on. Predefined attributes can be both intrinsic attributes, such as the creation date, and external attributes, such as the author: in such case whenever a reference to an artifact is inserted in RSAL such attributes must be filled and it is up to RSAL to keep them properly updated.

Attributes that are not predefined are called added.

Usual artifact descriptors such as faces, keywords, free text descriptions are regarded as attributes of an artifact.

Artifacts are connected by relations. A relation can be any kind of association that can be drawn between two artifacts such as one being the requirement of the other or one being the next version of the other and so on. Clearly, the definition of what is an artifact and what is a relation is not sharp: relations can rise to the level of artifacts when they have a self standing relevance, e.g. a baseline is a relation among several modules of code identifying a working product, therefore it is also a well defined artifact. Simplifying, RSAL may be thought as the catalogue of a library (a collection of cards describing in detail each book available in the library), and the artifact repository as the library itself. Before looking for a book in the library shelves, it is advisable to consult the catalogue. This is the principle adopted by RSAL. Each descriptor gives an exhaustive set of information for enabling the user in deciding if the artifact exhibits the characteristics he is looking for. The retrieval sub-system provides the user with powerful facilities for browsing the descriptor repository. Once the user has located the desired artifact descriptor, access to the artifact itself is provided by means of the tool used to create it or an equivalent one (supported tools).

## 3.1 RSAL architecture

RSAL offers an environment in which several users, working at their own workstations in different contexts, share the same pool of information. RSAL users see it as a central library whose retrieval facilities provide support in finding matches for all kinds of reusable artifacts stored in the Sodalia's software repository.

RSAL organizes artifacts of several kinds under a uniform view. To do that, it manages references or pointers to artifacts, rather than the artifacts themselves. Attributes and relations are the primary parts of an artifact reference. RSAL stores the references to the artifacts in a meta-data repository called the descriptors repository. This repository is directly accessed by the main modules of the system in order to archive, search, remove, change or retrieve an artifact descriptor. A Classification module is in charge of organizing the information about artifacts in a way that allows an efficient retrieval. An Identification module acts as a search engine on the descriptors repository, identifying the artifacts which satisfy given properties. This module is coupled and integrated with an interpreter handling SSQL queries -the SSQL module. RSAL produces some statistics on accesses to artifacts, and embeds an Announcement mechanism to inform end-users about new artifact acquisitions, new artifact versions, discovered bugs, etc. RSAL also performs application management functions to add users, set user privileges by category, and configure system parameters (e.g., keywords list). These last two functionalities are supported by the Application module.

A Tool Integration Platform underlying the whole system allows the interfacing with end-users and with application administrators, together with external tools, such as document tools or object oriented analysis and design tools.

Putting together these considerations, the RSAL high level architecture can be displayed as shown in figure 1.

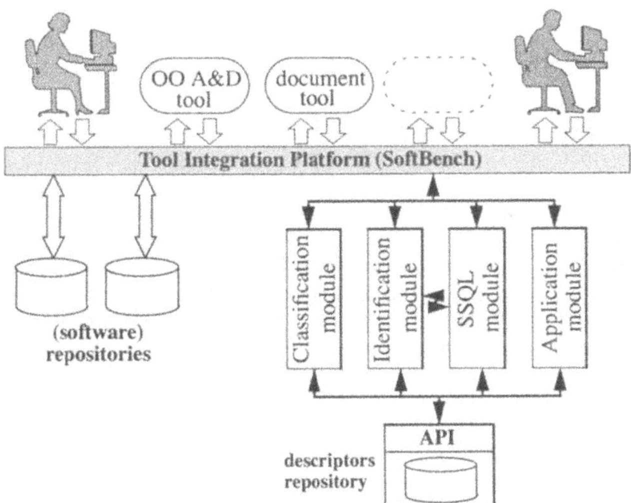


Figure 1: The RSAL general architecture.

## 3.2 Classification and retrieval

Classification of an artifact when it is inserted in RSAL allows the organization of all the attributes of the artifacts controlled by RSAL so that they can be efficiently and effectively identified and retrieved by the system. Effective reuse of existing artifacts requires a sophisticated classification/retrieval method. In the early stages of the reuse program, simple classification/retrieval methods are appropriate. However, more sophisticated ones are needed as the reuse library grows. Moreover, no single classification method is sufficient to find all relevant components of a given search (Frakes 1993b). Therefore, we consider multiple classification schemes, as explained below.

### *Classification of an artifact*

The three classification methodologies implemented in RSAL are different in the way the artifacts are classified and retrieved. When a new artifact is inserted, it has to be classified in all of the methodologies supported by the system. In fact, if this is not done, the artifact will not to be found when a search based on the missing methodology is performed.

For the **free text classification**, the artifacts are classified associating them with a text-description of arbitrary length which describes the artifacts themselves. For instance, a paper can be classified using its abstract.

In the **keywords classification** each artifact in the system is associated with a set of keywords that characterize the artifact itself and are chosen from a well defined dictionary. The dictionary can be expanded by the system administrator, who has the capability to remove unused keywords. For instance a class that implements a stack of integers could be classified using the keywords: integer, LIFO, source, C++, Object Oriented.

The classification of the artifacts in the **faceted** methodology is very similar to the keywords classification (Prieto 1989), (Sorumgard 1992). In fact, a set of keywords has to be associated each artifact in this case. The difference from the keywords classification is the dictionary from which the keywords are chosen: the dictionary is not plain, but structured into multiple facets that contain distinct subsets of keywords regarding different views of the artifacts. The single facets are structured as well. The keywords are stored in a weighted tree which represents their dependencies in terms of abstraction and specificity. The only user who can modify the facets is the system administrator. He can add or remove keywords, or move them to new locations in the trees; the system administrator can modify the weights in the trees, too, or even add or remove some facet. For instance if there are the facets Abstraction, Operates on and Dependencies, a stack of integers can be classified with the keyword Stack for the facet Abstraction, Integer for Operates on, and C++ for Dependencies.

### *Retrieval of an artifact*

An artifact stored in RSAL can be retrieved by navigating through the relations established between the artifacts or performing a search based upon one of the classification methodologies implemented in the system. The search is fundamental even when navigating through the relations - in fact, it is necessary to find an initial artifact from which to start the navigation. The search is done in different ways depending on the classification it uses: the user has to do a query in a format which depends on the different kinds of classification. If the results of the search are not satisfactory, the query can be relaxed or modified in some way and the search will be repeated with the new query. For the free text classification, the query consists of a string that describes the characteristics of the artifacts sought, and the system returns all the artifacts in which the textual description contains the search substring. The only way to relax the query is to change the string to look for. To use the keywords classification, the query is prformed by giving the system a set of implicitly ANDed. The system will return all the artifacts with all the keywords of the query present in their set of keywords. The query can be relaxed by specifying less keywords or by changing them.

Using the faceted classification, the user has to specify an arbitrary number of keywords to look for in each of the facets, and the artifacts at the minimum distance (calculated from the structure of the facets) from the query are returned. The query can be relaxed by changing the keywords to look for, or adding new ones. In fact, if multiple keywords are specified for a single facet, this is intended as an OR between them and the minimum distance between them is considered.

## 3.3 The RSAL query language

RSAL allows a user access to a metadata repository linked to one or more artifact repositories via predefined menus and windows. This kind of access to the data is very powerful but suffers a limitation: it is frozen and can not be changed to accomplish all the desired functions a user can imagine. One would prefer to have a system which is fully programmable by the user so that it can be configured to satisfy even very unusual problems. This can be performed using a special language which permits interacting directly with data in the repository to perform efficient searches.

RSAL introduces a flexible query language called **Set-based Query Language** (SSQL). The aim of this language is to allow a user to program his own functions in order to interrogate the repository. In particular, a repository is naturally viewed as a collection of (generally) homogenous data. A language especially efficient for treating sets (Jayaraman 1987) is thus a natural choice to begin constructing a specific query language. We based SSQL on a set-based language called SL, which is derived from the Subset Equational Language (Jayaraman 1992), (Jayaraman 1988). SL is very efficient in dealing with sets and it is quite simple to use. In order to interface this language with the specific kind of data used by RSAL, we have expanded the basic instructions of SL so as to allow direct query over metadata in the repository.

The extensions made to SL concern the management of the repository, introducing the notion of metadata and its fields. It is possible to refer to a particular metadata field in order to analyse all the object descriptions for a particular value. This has been achieved by introducing two classes of operators: a set of specific operators and a general wide-range operator.

The specific operators deal with predefined attributes, both intrinsic and external, which describe every artifact in the system. These specific operators follow the syntax:

Value = predefinedAttributeName(EntityName)

where Value is the value of the predefined attribute predefinedAttributeName for the artifact EntityName. For each attribute present in the artifact descriptor, a specific operator is provided to test the indicated metadata field. As an example for the private data of the artifact, we have the operators: author, creationDate, type, location.

Also, a general wide-range operator is provided. The syntax of this operator is as follows:

Value = test(AttributeName,EntityName)

with the usual meaning for Value, EntityName, making it possible to refer to the given AttributeName in order to test its value.

Note that the possible inconsistency arising from, for example, referring to an attribute not defined for the artifact indicated, is properly handled by RSAL. It returns an error code if the attribute has been never defined in the metadata repository or if the bottom value of the domain of all possible values if the considered artifact has not defined a value for it.

Implementing a search with a SSQL program is very straight-forward. For instance in order to retrieve all the artifact created by the user John before 31 May 1994, the program clause should be:

searchedEntities(Author) = {X : X in RSAL; author(X) == Author,
early(creationDate(X), (31,may,1994))}.

and the query should be:

searchedEntities('John').

Ending this paragraph, the following clause retrieves the author, the type and how many copies are present in the repository of all the entities that have the title specified in the head of the query:

searchTitle(Title) = {[Author, Type, Copies] : Artifact in RSAL;
title(Artifact) == Title; Author is author(Artifact),
Type is type(Artifact), Copies is test(Artifact, 'copies')}.

# 4 REUSE EXPERIENCES

Designing and constructing reusable components requires additional effort estimated from 5 to 10 times the effort required to build non-reusable software components (Meyer 1994). There is a definitive need for an organization that supports and fosters reuse activities, since project teams cannot be relied on to achieve significant reuse: developers are necessarily focused on their particular applications and are constrained in terms of time, costs, and resources. Thus, project teams will not want to spend additional time developing software that does not directly affect their own application (Caldiera 1991).

## 4.1 Reuse Support Organization

A Reuse Support Organization (RSO) is a group of software engineering experts (senior analysts, designers, and programmers with consolidated experience in object-oriented technology) that provide support to project teams in the following way:

- Training and education: RSO defines and enacts the policies to disseminate the reuse culture and to encourage the adoption of reuse practices among project teams. It is principally through this mechanism that the organization learns how to "reuse" effectively.
- Provide expertise: whenever possible, RSO engineers work with the project teams to better support them in achieving both development "for reuse" and "with reuse".
- Development of reusable components: dedicated teams are in charge of producing reusable components by either re-engineering the "normal" software components or by developing new components whose reuse potential has been assessed during domain analysis. The act of

developing reusable components asynchronously with project team activities is called an "a-posteriori (or backroom)" approach.

- Maintenance of reusable components: reusable components may change as flaws are corrected and enhancements are made.
- Maintenance and administration of RSAL: the central repository is managed by a single team which controls the nature and quality of the information stored in the repository. This activity also deals with the population of the library of artifact descriptors, including setting policies for artifact acquisition and maintenance (e.g., evaluation, certification, classification, and weeding). In addition it has to define and maintain a coherent classification scheme for the stored artifacts, weeding old artifacts, announcing incoming ones, user management (e.g., add/delete users, set user privileges), and reporting on artifact usage.

Nowadays, the RSO provides both expertise and support to the project team activities, and development and management of the reusable software components. Currently, development for reuse activity performed by project teams is limited to the discovering for opportunities of reusable components. This information is given to the RSO that adapts, by generalizing or reengineering, the normal (not the reusable) artifacts. However, project teams perform development with reuse activities looking for opportunities of reuse among already developed and classified components. These activities are based on the RSAL tool and are supported by the RSO.

## 5 CONCLUSIONS

Many organizations, commercial, academic, and governmental, are devoting resources to software reuse, although software reuse is not yet a major force in most corporate software development programs (Frakes 1993a). Successful reuse introduction into industrial organizations has demonstrated that reuse benefits, such as improved productivity and reduced time-to-market, are really accessible and lead directly to lower cost, higher quality software.

The Reusable Software Artifacts Library RSAL has shown the great importance of a reusable repository approach to leverage the efficiency and payoff of a corporate reuse program, especially in an environment where heterogeneous, complex and strictly correlated artifacts are potential candidates for reuse. RSAL plays a critical role in the introduction of a formal and systematic reuse program within Sodalia, allowing reuse and support information to be shared inside the organization. Its scope embraces both vertical and horizontal reuse.

Presently, it is relatively easy to expand the built-in functionalities of the system (e.g., to expand recognized artifact types and their associated tools), but this in itself could be made more automatic in the future, thereby quickly augmenting the rate of expansion of the system and the number of interested users. In addition, the system maintenance features should evolve. For instance, there should be a natural way of testing consistency if changes have been made to a remote system, and a way of notifying a remote administrator when such an inconsistency is detected. As well, many other event triggered notifications and actions are easily imagined. Finally, some feature for learning and automatic performance improvement are envisioned, perhaps being able to better guess what ultimate artifact a user is seeking based on his or her present search or browse actions and records of the outcomes of all similar such search or browse actions from past users.

A large-scale experimentation of RSAL has been set-up in the form of a project sponsored by the European Community under the ESPRIT program (project TARSAL #20555). This trial application aims to verify the use of RSAL in three different software companies with different reuse maturity, different software engineering culture, different application domains, different operating systems and different size.

## 6 REFERENCES


Caldiera, G. Basili, V. 1991 Identifying and Qualifying Reusable Software Components, IEEE Computer, February 1991.

Capers, J. (1994) Economics of Software Reuse, Computer, Vol. 27, #7.

Frakes, W. B. (1993a) Software Reuse Survey Report, Software Engineering Guild.

Frakes, W. B. (1993b) RSL System Concept Definition, Software Engineering Guild.

Jayaraman, B. and Plaisted, D. A. (1987) Functional Programming with Sets, Third International Conference on Functional Programming Languages and Computer Architecture, Portland, 194-210.

Jayaraman, B. (1988) Subset-logic Programming: Application and Implementation, 5th International Logic Programming Conference, Seattle, 843-858.

Jayaraman, B. (1992) Implementation of Subset Equational Erograms, Journal of Logic Programming, volume 13, number 3, 299-324.

Kain, J. Bradford (1994) Pragmatics of reuse in the enterprise, Object Magazine, 3(6), pp. 55-58.

Meyer, B. (1994) Library Design, Tutorial Notes, Tools Europe 1994, Versailles (France).

Prieto-Diaz, R. (1989) Classification of Reusable Modules, Software Reusability, Volume 1, Concepts and Models, T. J. Biggerstaff and A. J. Perlis publisher, ACM Press

Prieto-Diaz, R. (1993) Status Report: Software Reusability, IEEE Software, Vol. 11, #3.

Sorumgard, L. S. and Tryggeseth, E. (1992) Classification, Search and Retrieval of Reusable Software Components, Division of Computer Systems and Telematics, Norwegian Institute of Technology.

SSPG (1993) Sodalia Software Process Group, Strategic Product Architecture (SPA), Internal Document, Sodalia Software Development Infrastructure.

SSPG (1994a) Sodalia Software Process Group, SIMEP: Sodalia's Integrated Management and Engineering Process, Rel.3, Ver.1, Sodalia Software Development Infrastructure.

SSPG (1994b) Sodalia Software Process Group, SIMEP Software Metrics Model, Rel.1, Ver.1, Sodalia Software Development Infrastructure.

SSPG (1994c) Sodalia Software Process Group, Sodalia's Software Reuse Guidelines, Rel.3, Ver.1, Sodalia Software Development Infrastructure.

## 7 BIOGRAPHY

**Gianni Jacucci**

He was born in Rome in 1943.

He received the Laurea in Physics (cum laude) from the University of Rome in 1967.

He had a research fellowship of the Ministery of Education (1968-1971) and was a researcher of the Italian C.N.R. from 1971 to 1986.

He is a professor of computational physics at the Faculty of Engineering of the University of Trento Professor of computer science at the University of Trento, an adjunct professor of physics and supercomputing applications at the University of Illinois at Urbana-Champaign, a visiting professor at Computer Science Department, Cornell University, and at Computer Science Department and National Center for Supercomputing Applications, University of Illinois at Urbana-Champaign.

He has many scientific publications.

**Eliseo Mambella**

He was born in 1931.

He received the university degree in Sociology (cum laude) from the University of Rome.

He was employee of SIP (now Telecom Italia) from 1953 to 1991.

He is author of the Business Plan for the constitution of Telesoft and consultant for the managing of Telesoft from 1992 to 1993. He is a reviewer of the EUROWARE ESPRIT project.

He is the director of the Department of Research and Technologies of Sodalia S.p.A. since 1993.

His working interests focus on software engineering and include software reuse, reengineering, software process.

**Giancarlo Succi**

He was born in Vercelli (I) in 1964.

He received the Laurea in Ingegneria Elettronica (cum Laude) in 1988 and the Ph.D. in 1993, both from the University of Genova, a M.S. in Computer science from State University of New york at Buffalo in 1991.

He is a researcher at the University of Trento since 1993.

He organized two post-conference workshops at the "International Conference on Logic Programming 1993" in Budapest, Hungary and at the "Joint International Conference and Symposium on Logic Programming 1992" in Washington, DC.

He is author of several scientific publications.

His principal interests are software engineering and declarative languages.

**Carl Uhrik**

He was born in Cedar Rapids, Iowa (USA) in 1957.

He received a Bachelor of Sciences (Electronic) (B.Sc.) from University of Texas A&M, College station, USA in 1980, a Bachelor of Computer Sciences (B.Sc.) from University of Texas A&M, College station, USA in 1981, a Master of Sciences (M.Sc.) from University of Illinois, Champaign-Urbana, USA in 1986 and a Ph.D. in Computer Science from University of Illinois, Champaign-Urbana, USA in 1991.

Since 1990 he is a visiting researcher at University of Trento, Italy.

He was a visiting researcher at Computer Science Department and National Center for Supercomputing Applications, University of Illinois at Urbana-Champaign, USA, at National Center for Supercomputing Applications, Cornell University, USA.

He is author of many scientific publications.

**Marco Ronchetti**

He was born in Bolzano (I) in 1955.

He received the Laurea in Physics from the University of Trento in 1979.

He attended a Post-Doc at IBM-T.J.Watson Research from 1980 to 1981.

He is a researcher at the university of Trento since 1983.

He is author of 27 scientific publications.

He organized three workshops on scientific data graphic post-processing.

His recent interests are: simulation techniques in physics with supercomputers and workstations, distributed databases, image processing.

**Angela Lo Surdo**

She was born in Italy in 1953.

She received the Laurea in Mathematics from Rome University.

She worked for Italsiel S.p.A. (1978-1983) in development and maintenance of a TP monitor of IBM CICS, for Selesta Sistemi (1983-1990) on definition and coordination of the development and maintenance activities.

She was responsible for Configuration Management and System Administration from 1990 to 1991 and project leader in TCL domain area for Telesoft S.p.A. from 1991 to 1993.

Since 1994, she is responsible in Sodalia S.p.A. for Software Development and Project Management in the Research and Technology Department.

**Stéphane Doublait**

He was born June 27, 1963.

He received a Master of Sciences (M.Sc.) from Laval University, Quebec, Canada in 1990, a Bachelor of Sciences (B.Sc.) from Laval University, Quebec, Canada in 1986.

He was Principal Software Engineer at Bell Atlantic from 1990 to 1993.

He is Engineering Manager at Sodalia S.p.A since 1993, responsible for implementing the corporate reuse strategy of Sodalia.

He is author of more than 20 scientific publications in the area of artificial intelligence and software reuse.

His working interests focus on software engineering (software reuse, re-engineering, software process) and artificial intelligence (knowledge modelling & acquisition, automatic planning).

**Andrea Valerio**

He was born in Trento (I) in 1970.

He received the Laurea in Ingegneria Elettronica in 1995.

He is currently Ph.D. student at the University of Genova.

He is author of scientific publications.

His principal areas of interest are software engineering, especially software reuse, reuse metrics (both quality and productivity metrics) and declarative languages.

14

# Software design practice using two SCADA software packages.

*K.P. Basse ,G.K. Christensen and P.K. Frederiksen*
*Dept. of Control and Engineering Design*
*Technical University of Denmark*
*Building 421, DK-2800 Lyngby*
*Phone: +45 4525 4504, Fax: +45 4288 4024*

**Abstract**

Typical software development for manufacturing control is done either by specialists with considerable real-time programming experience or done by the adaptation of standard software packages for manufacturing control. This paper covers the experiences from the application of two software packages for Supervisory Control And Data Acquisition for the control of a chemical plant. The software packages are: "Fix" from Intellution and "InTouch" from Wonderware. Emphasis is placed on the practical software design cycle and the typical changes imposed on this by the application of the standard packages. It is argued that the SCADA and the PLC software must be developed in relation to a common specification for both applications although the development is done in two very different environments. Comparison of the software packages in relation to the process control specifications and the development effort is given.

Finally the application possibilities for these packages in relation to discrete parts manufacturing are evaluated based on the internal structure of the packages and the structural demands in this area.

**Keywords**

Supervisory control, standard SCADA software, software design.

## 1 INTRODUCTION

Ever since the introduction of the CIM-concept there has been a general consensus that the computer integrated factory is the factory of the future ( with a future !) . No matter what specialised mode of operation we are considering there has been increased efforts to use computers both for off-line planning and on-line production control. Inside the discrete parts manufacturing area the scene has been dominated by special purpose controllers like CNC-controllers, robot-controllers , more general process controllers PLC-controllers, etc. The process industries have relied more heavily on general process controllers and PLC's. Both areas face the challenge to integrate off-line planning with on-line production control. In general this challenge is expressed in terms like optimised production scheduling, quality control, product documentation and responsibility, CAD/CAM integration, etc. In both areas considerable integration has been demonstrated in terms of the functionality of the implemented plants. The demonstrations have mainly been achieved by completely new plants where one or two vendors have supplied everything from computer hardware and software to production machines or PLC's. The computer integration of manufacturing in companies with existing production facilities seems to go much slower due to the large investment in software development involved. One of the reasons for this is the lack of industrially accepted standards for the areas and the large amount of work involved in achieving commonly accepted goals. These goals may be summarised as ( Fix, 1994; InTouch, 1995; Mainstream, 1989):

* independence of hardware platform
* high level programming
* drivers to a large diversity of special hardware
* easy to set up graphical man/machine interface
* database access
* network capabilities
* transaction processing meeting real-time demands
* manufacturing specific support functions

On the software market there seem to be an increasing number of standard software packages available for manufacturing control. Some of these are termed SCADA software. SCADA is an abbreviation of : **S**upervisory **C**ontrol **A**nd **D**ata **A**cquisition. They are aimed primarily at various forms of process control and data acquisition tasks. Two successful packages that seem to fulfil the above mentioned general requirements were selected for the control of a typical chemical process plant. These were: Fix ( Fix,1994) and InTouch ( InTouch, 1995).The purpose was to test the packages in PC-versions against a specific set of process control specifications and to study the software design methods and related programming effort involved in the use of these packages.

## 2 THE SYSTEM SPECIFICATIONS

A simplified PI-diagram of the chemical chromatographic plant is shown in figure 1. Chromatography is a term used for a number of processes where different components of a mixture are separated in a process column by passing various chemical substances from input tanks through the column. The process does not demand real-time responses with reaction times of less than 1 second. In this respect the PC based software packages seemed not to present any problem. Nevertheless the PC is not used to directly control the plant. Because of high demands for control security and related pharmaceutical validation, the process has been controlled by a stand alone PLC .(SattControl,1995). The PLC contains the interface hardware for the process instrumentation and control valves. For the above mentioned reasons the structure for the plant control with SCADA software became a single PC/PLC configuration as shown in figure 2.
Additionally there was a demand that the PLC should be able to control the process also in the case where the PC might fail. This limits the SCADA software to the task of supervisory control and data acquisition. A short description of the control system specification is given below.

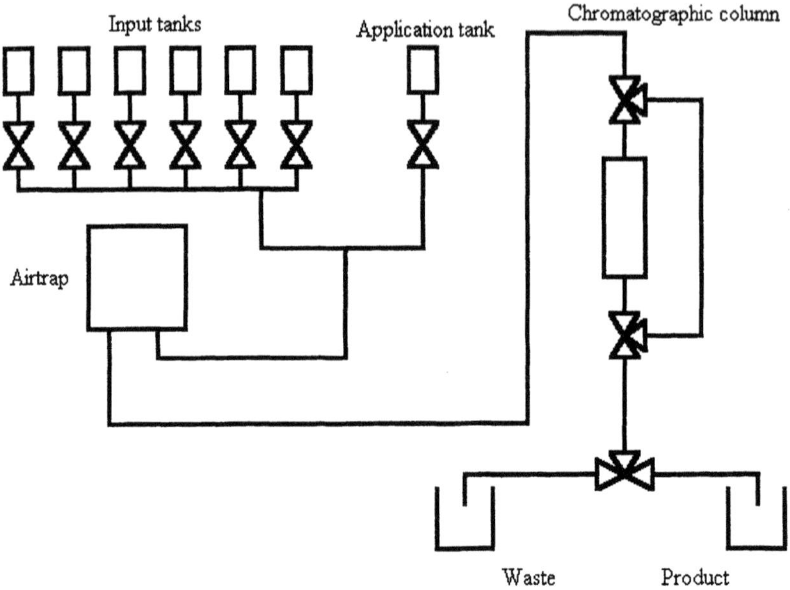


**Figure 1** Simplified PI-diagram for the chromatografic test plant.

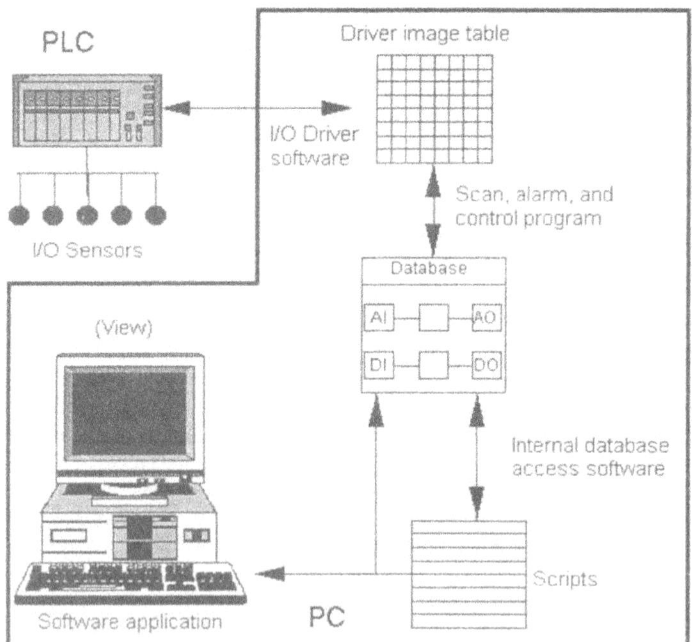


**Figure 2** PLC and PC interface structure.

The system will be in one of four states: Manual, Running, Idle or Error.

Manual: A state where the sequences of operations are all handled manually.

Running: A normal operating state where the PC/PLC -control system is running a number of batches. For each batch an initialisation is carried out, where the complete batch description is downloaded from the PC to the PLC.
Operator acknowledgements of alarms.
General data acquisition, logging and plotting should be done for two purposes, -one for on-line information on the PC-screen and one for historical documentation and off-line analysis of data stored in a database.
It should be possible to override the downloaded sequences by manual intervention to a certain degree, - for example to suspend normal running by entering an idle state.

Idle: A number of security actions are taken and further sequencing is halted.

Error: A state which only the system itself can generate according to a number of specific error conditions.

As can be seen from the foregoing description the system specification consists of a mixture of plant specifications and plant control specifications. The plant control related specifications are initially divided into a PLC-control specification and a SCADA- supervisory control specification for the control software design. In carrying out this division a specific effort is taken to clarify the interacting parts of the specification that is related to the communication between the controlling PC and PLC. This part of the specification is a result of the decision to divide the control task in a PLC and a PC-task. An important aspect is that any function required at the PC level that needs to access the process parameters can only get information or execute control insofar as these functions are supported by the PLC control software. Although the development was done by different programmers on the PC and the PLC level, it turned out that a close cooperation between these was necessary to allow a smooth transition into an integrated system. The advantage of using different programmers for the two control software tasks is that a higher degree of specialisation can be achieved. This is desirable taking into account the very different environments of programming and corresponding strategies that exist at the two levels. These methods and strategies will be dealt with in the following.

## 3 FUNCTIONALITY OF THE SCADA SOFTWARE PACKAGES

As the abbreviation SCADA indicates, the main objective of the software packages is Supervisory Control And Data Acquisition.
* *Data acquisition* is the ability to retrieve data from the plant floor and process this data into a useful form. Data can also be written to the plant floor, thereby establishing the critical two-way link that closed-loop control software require. The data transmission details of the communication with the process hardware are handled by an I/O-driver. The I/O-driver is selected for the hardware used in the plant, e.g. for the specific PLC used.
* *Supervisory control* is the ability to monitor real-time data coupled with the ability of operators to change set points and other key values directly from the computer. This function is typically achieved via a graphic interface which is constructed with the SCADA software packages in relation to the structure of the plant under control.
The tasks above are usually handled by the SCADA software package's central elements which are the graphical user interface, the database, and the I/O driver. A more detailed description of these elements is given in the following chapter, where the usage is illustrated by a simple example using the InTouch and Fix packages.

## 3.1 Other functions often found in Scada software packages

* *Data management* is the process where the acquired data is manipulated according to the requests of the software applications constructed by the user. The basic functions of data acquisition and management provide the basis for practically all the industrial automation tasks that the SCADA software packages can perform. The absolute *data integrity* is therefore an essential demand when automation is done with the help of these packages. Most of the products have some kind of tool to assist to ensure this integrity.
* *Monitoring* is the ability to display real-time plant floor data to operators. Powerful numeric, text, and graphical formats are usually available to make data more accessible.
* *Alarming* is the ability to recognise exceptional events and immediately report those events to the operators via the graphical display.
* *Control* is the ability to automatically apply algorithms that adjust process values and thereby maintain those values within predefined limits. Control goes one step beyond supervisory control by removing the need for human interaction. The computer can thereby be used to control the whole process or a part of the process to be automated.
* *Data archiving* is the ability to sample and store any data point in the system in data files at operator specified rates. At any time the data can likewise be retrieved from the data files to create trend displays of historical data. Managers and engineers can use this data to examine the events leading up to a critical event after addressing more immediate problems. The archived data represents a powerful tool for process correction and optimisation.
* *Reporting* is the possibility to access process data through industry standard data exchange protocols such as DDE(Dynamic Data Exchange) or ODBC SQL (Open DataBase Connectivity Structured Query Language). Operators can thereby create detailed reports with spreadsheets, compatible with MS Excel or MS Access, that contain acquired and calculated historical data.
* *Distributed processing* is the ability to create a system where the processing is distributed over a network.
* *Centralised processing* is the ability to have a system that contains only one computer (node) i.e. stand alone system
* *Time-based processing*

Most applications work by acquiring and calculating data at regular intervals, defined in seconds, minutes, or hours. Most SCADA software packages can perform any combination of *time-based processing*. This function allows the operator to balance the system resources giving priority to data that needs to be acquired quickly.

**Exception-based processing*

Processing that is triggered by events rather than time is known as *exception-based processing*. Processing can here be triggered by data changes, unsolicited messages from the process hardware, operator actions or other software applications. *Exception-based processing* is an essential feature for achieving distributed SCADA applications that monitor a large number of I/O devices.

# 4 TOOLS FOR PROGRAMMING THE SCADA APPLICATION

## 4.1 Introduction

During the construction of a SCADA application, the implementer encounters three different types of programming. These programming types are found when designing the graphical user interface, when setting up the database of the system, and when programming scripts. In the following sections these programming types are described in relation to Fix and InTouch, with some illustrative figures generated from a simple example. The example used is a 15000 litre horizontal tank where the volume in relation to the liquid level should be calculated and illustrated graphically. The input/output is volume/level or vice versa. The graphical user interface of this system is shown in figure 3.

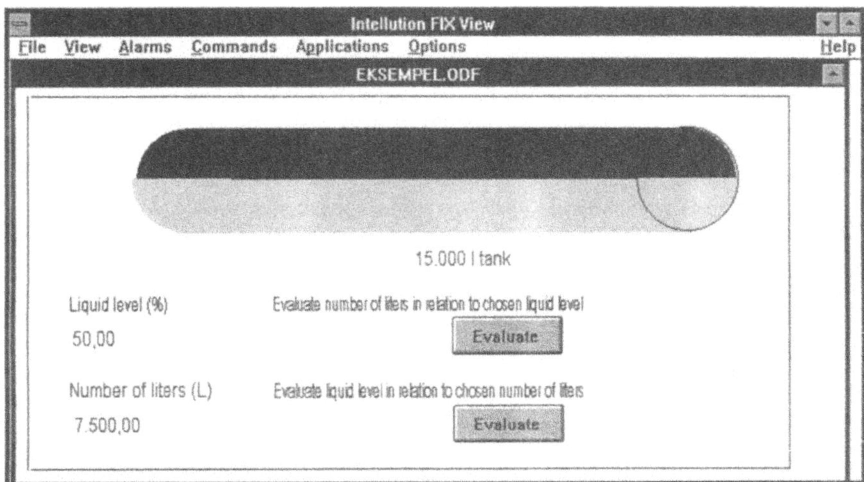


**Figure 3** Graphical interface for a horizontal tank. (Fix, 1994)

## 4.2 Graphical user interface

When constructing the graphical user interface, the SCADA packages provide some helpful tools for designing the graphic display. The most important is a toolbox where many possibilities for drawing and creating the computer displays, that allow the operator to interact with the process data are found. Another useful feature is the creation of smaller elements that are often used. These could be valves, tanks, etc. These elements are called Wizards in InTouch and Dynamoes in Fix. When designing the graphical user interface the constructor can re-use his set of predefined Wizards/Dynamoes and he will be asked to fill in the properties for each element. This is a powerful tool because it is possible to create a large library of standard elements that is very useful if there are strict demands for validation. The toolbox from Fix can be seen in figure 4.

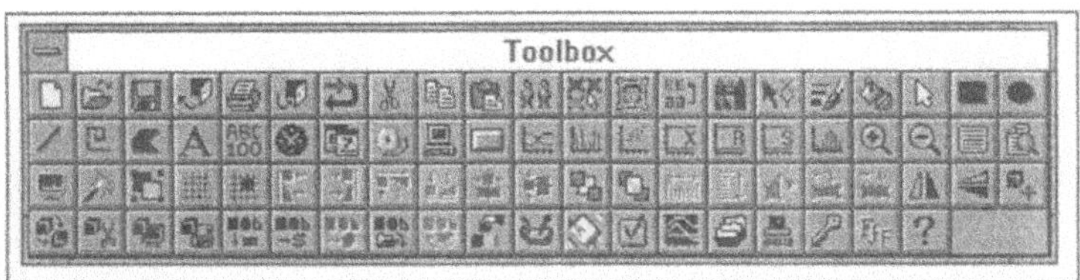


**Figure 4** Toolbox for programming via dynamoes (Fix, 1994)

## 4.3 Database programming

As mentioned above, the database is one of the central parts of the SCADA packages. The database of a SCADA package creates the connection between the process hardware via a driver to the graphical user interface via internal database access software as shown in figure 2. The first step in creating the database for the system to be controlled is configuring the driver that connects the database to the process hardware of the system that, as mentioned above, is typically a PLC. The driver is usually accessed by two different approaches - poll or event based, where Fix uses a poll-based driver and InTouch uses an event-based driver. A poll based driver continuously updates the data points whereas an event based driver only updates if there has been a significant change of the value in question. The implementer has to create driver image tables or poll tables where each I/O point from the PLC that is to be used in the database must be defined. Here the implementer must specify how often a data point is to be updated, which type of data is used, (whether is analog, digital or a register) and the specific address in the PLC. The drivers used by the two SCADA packages in question were not only different concerning the database updates. The driver used by Fix was a DOS program and the driver used by InTouch was a Windows program. As a result of this fact, Fix had to be restarted each time changes were made in the driver in order to acknowledge the changes, which was quite annoying. Once the driver image tables were completed the next step was to configure the database of the SCADA package. During configuration the main objective was to connect the driver image table to the graphical user interface. The first step was to decide what kind of variable, called a *tag*, was to be used. When this decision was made the implementer would be presented with a small window that is programmed via fill-in-the-blanks as seen for Fix in figure 5. From here it is possible to set up the hardware connection to the PLC, specify ranges and alarm values and so forth. When the database is fully equipped with all the connections to the hardware, the data that it contains is ready to be used in the graphical user interface direct or via scripts as explained in the following section. Values of the database can also be sent from the PC to the PLC. The Fix database builder for the example is shown in figure 6.

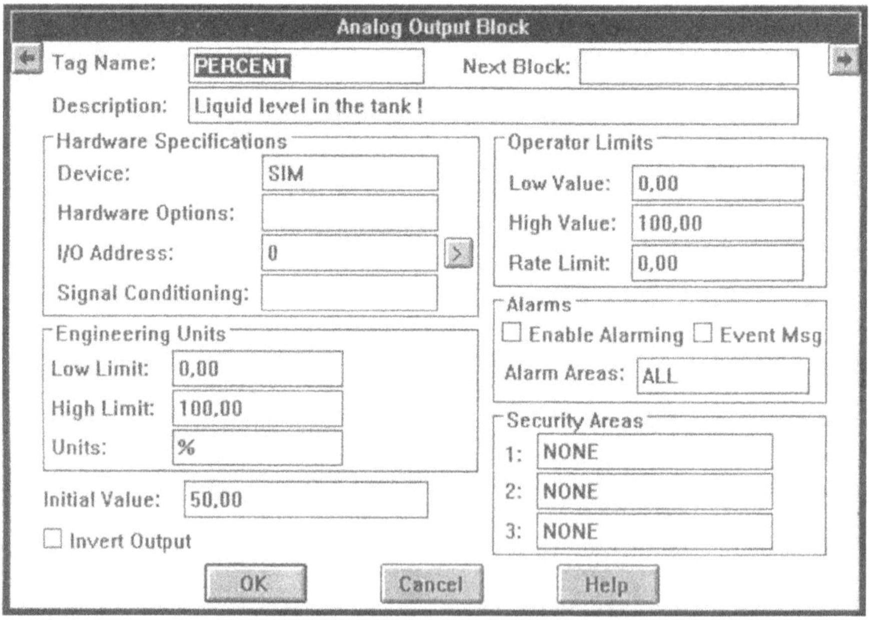

Analog Output Block
Tag Name: PERCENT
Next Block:
Description: Liquid level in the tank !
Hardware Specifications
Device: SIM
Hardware Options:
I/O Address: 0
Signal Conditioning:
Engineering Units
Low Limit: 0,00
High Limit: 100,00
Units: %
Initial Value: 50,00
Invert Output
Operator Limits
Low Value: 0,00
High Value: 100,00
Rate Limit: 0,00
Alarms
Enable Alarming
Event Msg
Alarm Areas: ALL
Security Areas
1: NONE
2: NONE
3: NONE
OK
Cancel
Help

**Figure 5** Fill-in-the-blanks programming (Fix).

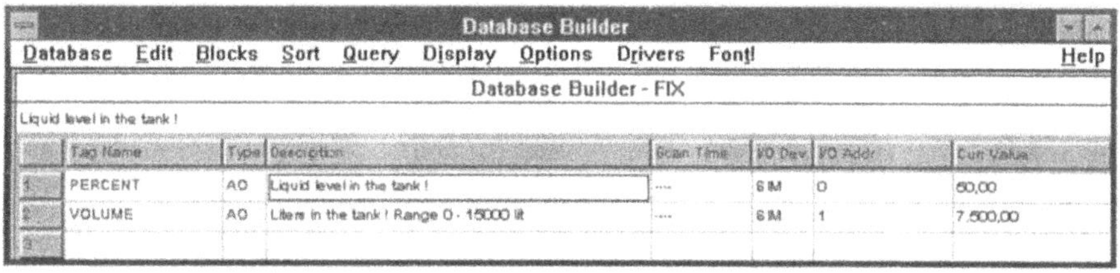

| | Tag Name | Type | Description | Scan Time | I/O Dev | I/O Addr | Curr Value |
|---|---|---|---|---|---|---|---|
| 1 | PERCENT | AO | Liquid level in the tank ! | ---- | SIM | 0 | 50,00 |
| 2 | VOLUME | AO | Liters in the tank ! Range 0 - 15000 lit | ---- | SIM | 1 | 7.500,00 |
| 3 | | | | | | | |

**Figure 6** Database builder (Fix).

## 4.4 Script programming

The final step in constructing the application is to use the database tags in scripts, which are small programs that often initiated by an operator. These programs can alternatively be started by events taking place in the application. The scripts are used for manipulating data from the hardware so it is possible to present these in just the right way in the graphical user interface. The programming tools for programming the scripts are for Fix and InTouch quite different. Fix uses a programming language, called Command Language, where most of the commands are specific for the Fix package. The programming language used by InTouch is more similar to standard languages like Fortran and Pascal. Another important difference is the size of the programs. Fix only has the possibility to make scripts containing 50 program lines, where InTouch support almost an infinite number of program lines. This makes script programming in Fix very difficult because space is very limited. Both of the mentioned SCADA packages have promised changes in this field, and future releases of the script language will be based on Visual Basic. Because the Fix package uses a special set of commands it is time-consuming for a novice to construct perfect scripts. In figure 7a and figure 7b the scripts used in Fix and InTouch to handle the examples above are shown to give an idea of the differences.

```
1 DECLARE #VOLUME NUMERIC SCRIPT
2 DECLARE #TEMPVOLUME NUMERIC SCRIPT
3 DECLARE #HEIGHT NUMERIC SCRIPT
4 DECLARE #PERCENT NUMERIC SCRIPT
5 DECLARE #I NUMERIC SCRIPT
6 GETVAL FIX:VOLUME.F_CV #VOLUME
7 #TEMPVOLUME = 7500
8 #PERCENT = 50
9 IF #I <= 20
10  GOTO 15
11 ENDIF
12 IF #I > 20
13  GOTO 29
14 ENDIF
15 IF #VOLUME == #TEMPVOLUME
16  GOTO 29
17 ENDIF
18 IF #VOLUME > #TEMPVOLUME
19  #PERCENT = #PERCENT + 50 / (2 ^ (#I + 1))
20  #I = #I + 1
21 ENDIF
22 IF #VOLUME < #TEMPVOLUME
23  #PERCENT = #PERCENT - 50 / (2 ^ (#I + 1))
24  #I = #I + 1
25 ENDIF
26 #HEIGHT = #PERCENT * 17,5
27 #TEMPVOLUME = 2 * 6236,2753 / 1000000
   * ((#HEIGHT / 2 - 437,5) * SQRT(-(#HEIGHT ^ 2)
   + 1750 * #HEIGHT) - 382812,5 *
   ASIN(1 - (2 * #HEIGHT) / 1750)
   + 191406,25 * 3,1425926535)
28 GOTO 9
29 SETVAL FIX:PERCENT.F_CV #PERCENT
```

**Figure 7a** Script example from Fix for the horizontal tank.

```
Volume = VolumeWindowValue;     {Volume first altered when "OK"}
TemporaryVolume = 7500;          {First guess, tank half full}
TemporaryPercent = 50;
FOR i = 0 TO 20          {Max. 20 iterative actions}
IF Volume == Round(TemporaryVolume,1) THEN
   EXIT FOR;   {Exit if calculated volume equals the volumen given}
ENDIF;
IF Volume > TemporaryVolume THEN
{If the calculated volume is smaller than the given volume}
   TemporaryPercent = TemporaryPercent + 50/(2**(i+1));
   {New guess is calculated}
ELSE
   {If the calculated volume is bigger than the given volume}
   TemporaryPercent = TemporaryPercent - 50/(2**(i+1));
   {New guess is calculated}
ENDIF;
Height = TemporaryPercent * 17.5; {Height of tank is 1750mm}
TemporaryVolume =
 {Calculation of volume corresponding to new TemporaryPercent}
 2*6236.2753/1000000*(((Height)/2-437.5)*Sqrt(-(Height)*(Height)+1750*(Height))
 -382812.5*ArcSin(1-2*(Height)/1750)*3.1415926535/180+191406.25*
 3.1415926535);
NEXT;
Percent = Round(TemporaryPercent,1);
Hide "Pop-up window for volume";
```

**Figure 7b** Script example from Intouch for the horizontal tank.

## 4.5 Application programming practice

When programming a new application with the assistance of SCADA packages, it was expected that the approach would vary with the amount of programming experience of the programmer. An experienced programmer will often find the top - down/bottom - up principle quite helpful. This method results in an approach where the total specifications for automation of the system are generated according to the top-down principle mentioned above. Afterwards the actual programming can begin. The code writing then follows the bottom-up principle, where programming starts at the PLC. Once the PLC program is completed, the systems I/O driver is configured. This is done by defining all the I/O points to be used in the application. The next step in the process is to configure the database of the SCADA package according to the I/O points defined in the driver, and the needs represented by the graphical user interface. When these tasks are completed, the graphical user interface is designed in relation to the wishes and demands presented in the specifications. The user screens are mostly constructed with the aim to have some kind of resemblance with the physical system to be controlled. The final process is to connect the elements of the graphical user interface with the database directly or via programming scripts. At this stage the dynamics of the system are made accessible to the operator. The approach presented above is an ideal approach which for beginners was found quite hard to follow. During the programming it was experienced that it was often necessary to move back and forth between the levels as the experience grew. This movement was however not felt by the programmers as a difficulty, because the path from implementation to test was assisted greatly by the software packages. A useful approach here was to advance the graphic presentation part of the software in order to present the planned functionality and to use it as a basis for discussion with the users.

# 5 TOOLS FOR PROGRAMMING THE PLC

A number of methods for creating PLC-code were identified: Instruction List (IL), Ladder-Diagrams (LD), Function Block Diagrams (FBD) and Sequential Function Charts. (SFC), (IEC 1131-3). What we needed was a high level language and graphical presentation that was easy to present and discuss with plant users. We chose the SFC method, sometimes referred to as GRAFCET (B. Goran, 1991) which is a graphical method that is excellent for the sequential part of the control, but does not describe data management aspects of the PLC program. The PLC selected for the process was a SattControl 05-45 PLC with 65 Kbytes of memory and could be programmed by a PC software package: DOX-5/10. Because an old version of the software has been written using DOX-5, this version was used. The PLC is equipped with 64 digital I/O points, 8 analog inputs and 4 analog outputs. Even though DOX-5 did not support GRAFCET programming, the method was used throughout to discuss specifications with users and as a specification for the programmer. An example of a list of sequential steps using GRAFCET is shown in figure 8.

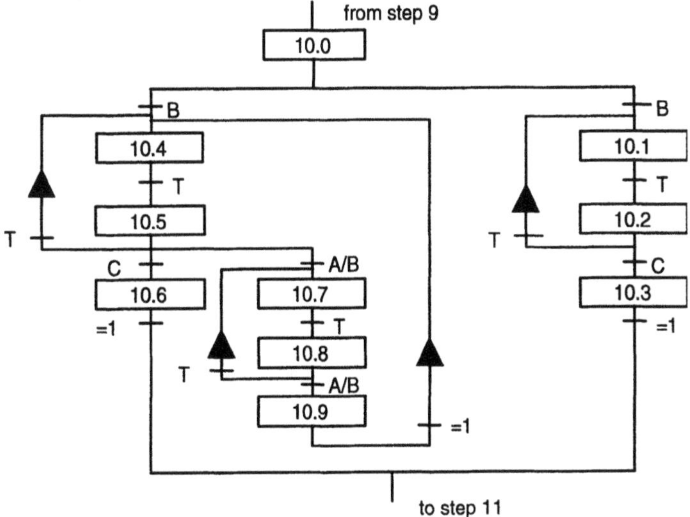


**Figure 8** Example of a Sequential Function Chart from the test plant (GRAFCET)

The sequential function chart specifies the central part of the control software, and the other parts are more or less derived from, or merge into, ist states. The output does not normally change without an initial change of state and the states are the basis for forming the output. As additional guidelines for programming, a set of company and PLC specific function blocks and guidelines were used. The function blocks dealt with: self-locking states, sequential steps, component controls and calculations. The guidelines for programming addressed naming conventions, structuring of code and other hints for writing good, serviceable software code. Despite the attempts to keep programming at a standard level this was not totally possible with the PLC-code. Like most PLC applications, the coding was sometimes rather awkward and vendor specific when we got beyond the basic Boolean operations programming. An example is a case where a user demand results in a corresponding demand for flexibility of the PLC program. In the test plant the user wanted to be able to change the set-up of input and output tanks from one GRAFCET state to the next (see figure 1). The solutions suggested were:

a) download a complete PLC-program containing the new input/output specification

b) download a new input/output specification for each change of state

c) download a complete input/output specification for all states

The first alternative conflicts with the desire to keep validation at a minimum, using only one code variant. The second conflicts with the specification in saying that the PLC should be capable of controlling the process in a stand-alone mode. The third method is the only one acceptable. It requires however a set of register and addressing capabilities that would be easy to handle in, for example PASCAL or C, but requires a PLC-code specialist using, e.g. indirect addressing. The result is that a lot of vendor-specific programming tricks are required and the programming for just this vendor product line becomes both time consuming and costly. This again led to a company policy to limit the number of different PLC vendors in order to keep programming and maintenance problems at a manageable level.

## 6 EXPERIENCES GAINED

In 12 months two beginners of both SCADA and PLC-programming and with only basic PASCAL programming experience set up specifications for and implemented one PLC-program and two SCADA solutions using FIX and InTouch. The specification was a renewed specification for the complete PC/PLC-solution based in part on an existing PLC-specification and program. The new specification took about 6 months in close cooperation with process and plant engineers and operating personnel. The specification process was as usual an interactive process trying to assemble loose ends, harmonise different views and scale ambitions. In the same period, the students became acquainted with the software products at hand: The PLC-programming environment and the two SCADA products. After another 6 months one complete, working process control system was demonstrated and the accompanying documentation finished. The effort is described as 50 % PLC-programming and 50 % SCADA programming.

During implementation some minor problems with the SCADA products were found. The FIX package did not allow for a specified automatic lock-out in the case of inactivity. Concerning the database interface some minor problems were detected when trying to export data from InTouch to the Microsoft database Access. Furthermore it was felt that the SCRIPT language for FIX was less intuitive than the more general purpose SCRIPT language for InTouch. Summarising, the SCADA packages had several ways to facilitate high level programming: Fill-in-the-blanks, wizards-programming and clip arts for the graphics interface. However troublesome these programming tools were felt by the programmers, there is no way they could have made the same kind of control system without this extensive programming support.

Moreover the features of the packages that were not tested or used presently for the process control are still available for future extension and integration in the company. Thus path to "open" systems using de facto standards like Windows, DDE, NetDDE and ODBC SQL achievable. One important aspect of programming was not carried out during the above mentioned period - the final system validation. The validation procedures followed by the company is regulated to a large extent by the pharmaceutical regulations. The validation process was estimated to take an additional 3-6 months. (R. Konakovsky, 1994). The uncertainties relating to the validation of the large software package running under Windows further underlined the necessity for a solution where the critical parts of process control could be run by the PLC alone.

## 7 CRITICAL ISSUES

The application of standard software speeded up the development of the application, but also restricted the freedom and insight of the programmer. In this case, as in many real-time control applications, it is almost impossible for the programmer to estimate the time required in data acquisition, data management and data presentation. In this case no extreme demands for event management and small scan-times were specified. The fastest scans were measured in seconds and minutes. Concluding, it would be very useful to have a set of standard tests or benchmarks for different configurations of SCADA products and PLC's. In this respect, special attention must be given to the driver and communication hardware.

## 8 FUTURE ASPECTS

The software development described leads to the conclusion that considerable development time can be saved by integration of the two software tasks for the PLC and the SCADA application. The integration is illustrated in figure 9. Here the PC and the PLC melt into a “SoftPLC”. The advantage is that the construction of the PLC-code can automatically build the database for the SCADA application. Furthermore, the interaction between the PLC and SCADA functions are optimised, not dependent on different drivers. Another feature illustrated by figure 9 is the possible change of the process instrumentation interface, which can be shifted towards a fieldbus solution, where all sensor signals are sent via the network in digital form.

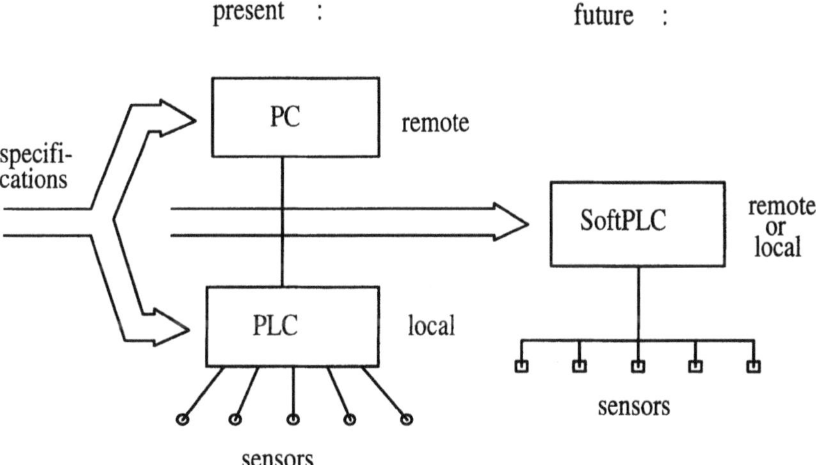


**Figure 9** Future developments related to PC and PLC-programming.

## 9 APPLICATION TO DISCRETE PARTS MANUFACTURING (DPM)

The software packages for SCADA applications are attractive for DPM because they represent a large set of features also requested for discrete parts manufacturing. Additionally the packages are in widespread use for control tasks in the process oriented industries.

Two specific issues come up in this regard. One is the question of communication with CNCs, AGVs and Robots. A list of general communication requirements for this area is shown in table 1. The SCADA products can probably handle the communication requirements via device specific drivers written by experienced C-language programmers for the non standard manufacturing equipment at hand. The main issue is whether the SCADA-products can support the requirements for the manufacturing of discrete parts. In DPM there is a need, not only to monitor and control the state of the plant, but additionally to keep track of each individual part. This requires the SCADA product to be able to handle several plans simultaneously, one for each part. These, and other specific requirements that could make SCADA products much more interesting for flexible DPM are a topic for future investigations.

**1) upload and download of programs to a device**
**2) dynamic downloading**
**3) notify supervisors**
**4) start and stop of programs**
**5) ask for status**
**6) change data on a device while executing a program**
**7) control auxiliary equipment**

**Table 1: Communication functionality for DPM -equipment (G.K. Christensen,1992)**

## 10 REFERENCES

Fix Dmacs ver.5.0, System Setup, System Development, Advanced Tools and Display Development, Intellution, Inc. 1992-1994.

InTouch, ver.5.0 -manuals, Wonderware, Inc. 1995

MainStream- Application Integration Platform for Open Factory Systems, ITP Enterprice Software Inc, Cambridge, Ma, 1989.

SattCon 05 Slimline (V3) - Installation and maintenance, Alfa Laval Automation A/S, may 1994, Dok. nr. 493-0573-01 version 1.1.

DOX-5 Users manual, SattControl, AB Malmo, november 1990., Art nr. 493-0235-11.

B.Goran og G.Sandberg: "Funktionsdiagram - et kursushæfte til beskrivelse af SS IEC 848, S -konsult AB, ISBN 91-87-18220-3.

J.H. Christensen: "Function Block Standardization - Liason Report, ISO/IEC, Presented to ISO TC184 Meeting, Torino, Italy, 17/5 1995.

IEC 1131-3 (PLC Languages) IEC 1992.

G. K. Christensen and C. Nøkleby: "Quality Interfaces In The 90ties Using MMS", Systec, MAP/MMS workshop 3, 1992, Munich, Germany.

Rudolf Kanakovsky and Peter Woitzik: "Automatisierung der Typprufung von Software fur Prozessleitsysteme", Automatisierungstechnische Praxis 36 (1994). 10 , 12-21

# 15

# GENIUS: A Generator for Graphical User Interfaces

*Univ.-Prof. Dr.-Ing. Prof. e. h. Dr. h. c. Hans-Jörg Bullinger*
*Dr.-Ing. Dipl.-Math. Klaus-Peter Fähnrich*
*Dr.-Ing. Dipl.-Inform. Anette Weisbecker*
*Fraunhofer-Institut für Arbeitswirtschaft und Organisation (IAO)*
*Nobelstr. 12c, D-70569 Stuttgart, Germany, +49 711 970 2320, +49 711 970 2300, Anette.Weisbecker@iao.fhg.de*

**Abstract**

GENIUS (GENerator for user Interfaces Using Software ergonomic rules) is a system that generates ergonomically designed graphical user interfaces from extended data models. GENIUS is based on a methodology to develop graphical user interfaces in a stepwise procedure starting from standard data models of a specific application system (e.g. production planning and control systems). The GENIUS system transforms these models in a series of consecutive operations into an operational graphical user interface (GUI).

The methodology introduces the definition of views for the specification of the data and functions required by the users. In addition, the derivation of the dialogue structure from the data model is presented. Based on the defined views, the automatic generation of the user interface is carried out by a rule-based system using explicit design rules derived from existing guidelines. Output is generated for an existing user interface management system. The approach supports rapid prototyping while using the advantages of standard software engineering methods and ensures the integration of these methods with user interface design.

GENIUS has been tested for the development of a series of graphical user interfaces for production planning and control systems. It has been used in typical migration situations from alphanumerical systems to be migrated into client/server solutions with graphical user interfaces as well as for the development of new systems for different platforms.

**Keywords**

Automatic User Interface Generation, Data Models, User Interface Management Systems, Production Planning and Control System

# 1 INTRODUCTION

Data and information have become important economic factors and major milestones for industry. Therefore, systems are required within a company for integrated and constant information processing. These systems must support the fulfilment of competitive objectives such as cost reduction, processing time minimization and improvement of quality. This is especially true for customer-oriented information systems within a company which are indispensable for an efficient processing of customer orders.

There is evidence that the functionality of information systems within a company is not fully used (VDMA, 1992). The provided information appears as a huge amount of data the users cannot cope with. Nonetheless it is essential for the successful use of information systems, that the required information is represented in a comprehensive way to the user. This means that the application of software ergonomic design guidelines is imperative during software development. At present the development of software ergonomically designed user interfaces is still time-consuming and often very difficult for the system developers, because they lack the appropriate knowledge (Hüttner et al., 1992).

Although a greater number of user interface development tools is available, it is difficult to follow existing user interface design guidelines and style guides (Tetzlaff, Schwartz, 1991), because the tools do not provide support for software ergonomic guidelines. In addition, user interface tools cannot make use of the models developed with general software engineering methods and tools in the early phases of the design cycle, which specifies the non-interactive part of an application. The integration of application development and user interface design which has been often demanded but is still missing in practice, leads to additional effort and potential inconsistencies when moving from requirements analysis to user interface design (Foley, 1991). It is therefore important to use analysis results such as the data model as input for user interface development tools.

The work described in this paper addresses the problems mentioned above. It shows a method and the supporting tool environment GENIUS (GENerator for user Interfaces Using Software ergonomic rules) for the automatic generation of user interfaces from extended data models (Weisbecker, 1995) and the use of GENIUS for the development of a graphical user interface for a production planning and control system. GENIUS generates user interfaces according to human factor guidelines and guarantees the integration of software engineering and user interface design by using the application data model as a basis for application development as well as for user interface design.

The development of graphical user interfaces with GENIUS is carried out in three main steps (Figure 1), which are described in the following chapters. The first step is interactively performed by the software developer and the other two automatically by GENIUS.

- **Specification**
  In the first step the data and functions necessary for a task are specified. This is carried out by defining the views according to the user's task and on the basis of the data model in the form of an entity relationship model.
- **Generation**
  The information specified in the first step is used for deriving the dialogue structure. In the second step it serves as the basis for generating user interfaces while applying software ergonomic rules.

- **Implementation**
  In the third and last step the generated user interface description is transformed into the specification of a user interface management system (UIMS), which is then responsible for the implementation of the user interface. This last step permits the use of different user interface tools together in combination with GENIUS.

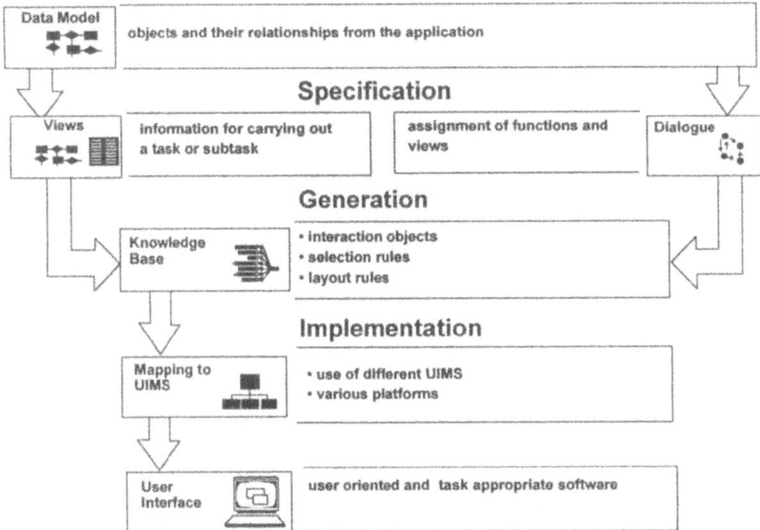


**Figure 1** Overview on GENIUS environment. Views are defined based on the data model. A rule-based component generates the user interface presentation and the dialogue structure according to the view definitions. As a target system, an existing UIMS is used.

## 2 RELATED WORK

A number of research systems are documented in the literature that automatically generate the user interface from higher level specifications. Jade (Myers, vander Zanden, 1990) and ITS (Wiecha et al., 1990) are generating presentational designs from frame-like dialogue specifications. Within the UofA* (Singh, Green, 1991) the component Chisel (Singh, Green, 1989) is used to generate presentation and dialogue of application commands by means of design rules with regard to user's preferences and given device characteristics. Mickey (Olsen, 1989) uses the declarative constructs of the programming language Pascal for the description and generation of Macintosh user interfaces. UIDE (Foley et al. 1991), in combination with DON (Foley, Kim, 1990) uses a specification consisting of objects, attributes, attribute types, actions, parameters, preconditions and post conditions. Each of these systems introduces its own notation for the higher level user interface specification. However, the use of widely known notations is important for the acceptance of tools by system developers.

HIGGENS (Hudson, King, 1987) is the first UIMS which incorporates data models and uses views to represent an abstract form of the display. In contrast to the GENIUS approach and most of the systems mentioned above, HIGGENS does not incorporate rules for the generation of the physical user interface. Thus, it offers less flexibility and power with respect to the selection and layout of interaction techniques.

The system described by de Baar, Foley and Mullet (1992) generates dialogue boxes and menus from enhanced data models. With this approach, descriptions from data modelling can be reused for the user interface specification, and double effort and consistency problems are avoided.

In the approach of Petoud and Pigneur (1990), the entity relationship models are used for user interface generation. The dynamics are textually specified and can be visualized by a precedence graph. The graph, however, can neither be modified nor substructured. Therefore, the system is probably not suitable for large applications.

As a whole, existing systems for the automatic generation of user interfaces either lack a proper integration into general software engineering methods, or sufficient support for user interface design guidelines.

## 3 USER INTERFACE SPECIFICATION BY VIEW DEFINITION

### 3.1 Deriving User Interface Presentation from Data Models

The data model in the form of an entity relationship model is the starting point for the specification of the user interface with GENIUS. The entity relationship model (Chen, 1976) is a well established conceptual data model. It is used in most of today's software-engineering methods (Yourdon, 1989) and is also the most frequently used modelling technique in industry (Bittner et al., 1992). In addition, it is supported by a great number of computer-aided software engineering (CASE) tools.

The entity relationship model shows the elements of the application area and their relations; nonetheless, it does not describe the data the user need for performing specific tasks. Therefore so-called views are defined. A *view* consists of a subset of entities, relationships, and attributes of the overall data model. Additionally functions are assigned to a view, thus a view comprises the whole information to perform a specific user task. Within these functions two different types are distinguished: data manipulation functions and navigation functions. Data manipulation functions can be applied to data elements in the view. Navigation functions determine the dialogue structure by calling other views. They are used to define the dynamic behaviour of the user interface.

In order to enable an easy and efficient view definition, a number of extensions have been added to the entity relationship model. For specifying logical groups of attributes, complex attributes are introduced. The grouping of information is used in the user interface generation process for visualizing groups of related data. For reducing the complexity of the graphical representation of entity relationship diagrams, a set of entities and relationships can be replaced by a single symbol and edited separately.

Within the GENIUS tool environment, the definition of views are defined by a direct manipulation editor. The application developer interactively selects elements from the entity relationship diagram. Thus, no programming skills are required for the definition of the views, and it is easy to involve non-programmers into the process e.g. application experts and interface designers.

Because the entire information needed for the generation of user interface windows from views cannot be expressed in a graphical representation, it is necessary to supply additional textual descriptions. These are provided in so-called *property sheets*. Property sheets can be popped up during the definition of the views and edited directly. They consist of a definition of each element which in most cases is available from the data dictionary, the assignment of the functions to the view and the inclusion of task oriented properties which are used to map the application elements on appropriate interaction objects in the automatic generation process.

According to the elements in a view and their structure, two different types of views are distinguished:

- Aggregation View
  An aggregation view shows a collection of different elements e.g. entities, relationships and attributes. According to the presentation of the aggregation view two different forms have been characterized:
  - Reference Presentation
    A reference presentation shows the aggregation as a whole. It may also be called a container. Usually, entity types are represented by containers. Containers are the typical elements for the first dialogue steps (e.g. in Figure 5 parts of a production planning and control system).
  - Set Presentation
    In a set presentation each element of an aggregation is shown. Depending on the number and structure of the elements, lists, sets or other presentation structures such as trees are used. The elements shown in the set presentation give references to the detail view of each element.
- Detail View
  A detail view shows the single attributes of entities. Either all or only some attributes of one entity may be shown, but also attributes of different entities which are needed in regard to a specific task (Figure 7).

The structures in the data model are preserved in the views. They are used to determine the presentation of the attributes and to derive the dialogue structure.

Within the automatic generation of the user interface, the view elements are mapped to appropriate interaction objects by means of selection rules which take into consideration the characteristics of the attribute such as type, range and number of values. The relationships between the entities also influence the presentation and structuring of the view elements. For example, the attributes belonging to the same entity or relation are implicitly grouped in order to reflect their logical connection to the user. In addition, the cardinality of relationships between the entities influences the presentation of the attributes. Thus in the case of a 1:n or n:m relationship, an aggregated view is used to show the different instances.

## 3.2 Deriving Dialogue Structures from Views

The structure of the views does not only serve for determining the presentation, it also supports the automatic derivation of the dialogue structure, because both object-oriented user interface and data model are structured according to the data objects. Therefore, the data model structure can be transferred to the dialogue structure for an object-oriented user interface.

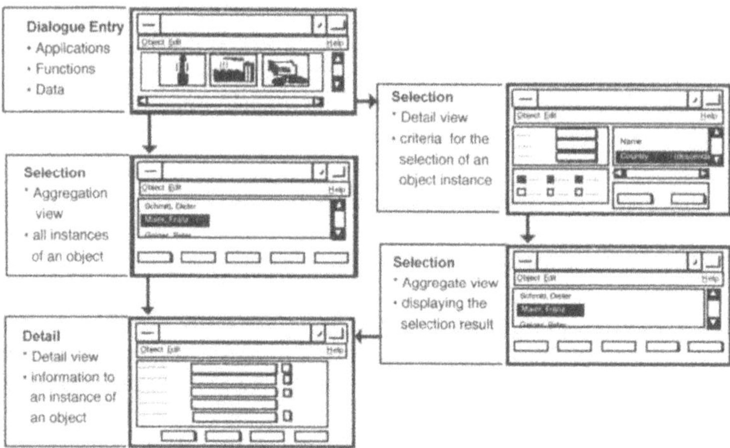


**Figure 2** Schematic dialogue structure for object-oriented user interface.

The dialogue structure of an object-oriented user interface consists of three main parts. The first part is the application entry, which shows all relevant objects. In the second part a selection dialogue is provided in order to identify the desired object. The last part shows detailed information on the desired object. Thus, in a typical object-oriented dialogue, navigation to the desired object must be carried out first in order to perform a specific task in the second step.

The starting point for the definition of such a dialogue sequence is the coarse data model which contains merely entities and relationships (Figure 4). Based on this model, container views can be defined in order to provide a dialogue entry (Figure 5). A container view consists of those entity types which do not result from normalization.

After selecting a container object, a possibility for further selection of the desired object instance must be provided. Depending on the number of instances, three different ways can be used. If the number of instances is small, icons can be used to show the single instances of the object. For a medium-sized number of instance, a list is appropriate. A search form with criteria for further specification is provided if a large number of instances is available (Figure 7). The search process results in a list showing the objects which fulfil the search criteria. The selection of an entry in the list will lead to the details of the desired object.

The aggregation and detail views needed for such dialogue sequences can be defined interactively on the basis of the data model; to some degree, they can be automatically generated. In the generation process a window will be assigned to the container view and the entity types will be represented by icons. Depending on the number of instances for each entity type an aggregate view will be generated or a detail view for a search form is added. If the aggregate view or the detail view for a search form is generated automatically, either the key attributes of the entity will be shown or the attributes which are explicitly marked as identifier for selection will be used.

# 4 GENERATING THE USER INTERFACE

The defined views form a logical description of the application user interface. This description is used as input for a rule-based system carrying out the generation process. The generation is performed in three steps (Figure 3). In the first step appropriate interaction objects are chosen for the presentation of the data and functions specified in the views. In the second step the parameters (e.g. colour, position, size) for these interaction objects are determined. In the third step the layout is done by arranging the interaction objects.

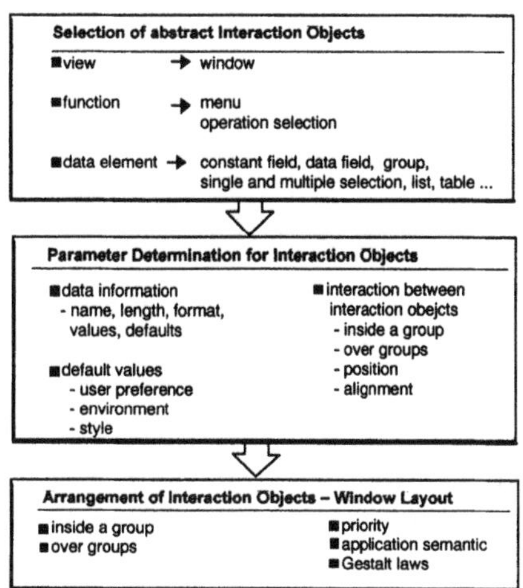


**Figure 3** Main generation steps in GENIUS.

## 4.1 Mapping View Elements on Abstract Interaction Objects

For the first generation step *abstract interaction objects* have been introduced. These interaction objects are called abstract because they are independent of their physical implementation, e.g. as MS-Windows, OSF/Motif or OS/2 Presentation Manager widgets. Thereby, flexibility with regard to the target environment is reached and the portability of the application is supported. The definition of the abstract interaction objects is based on an examination of available interaction objects from different style guides (OSF, 1994; IBM, 1992; Sun, 1990; Microsoft, 1992). Standard interaction objects defined for graphical user interfaces are windows, dialogue boxes, menus, entry fields, constant fields, exclusive choices, non-exclusive choices, operation choices, lists, and scroll bars. They are called abstract because they are independent of their physical implementation, e.g. as OSF/Motif or Open Look widgets.

The mapping of the defined views, functions and attributes on the appropriate abstract interaction objects is carried out by selection rules.

These rules are derived from existing guidelines and style guides. Most of the rules concerning formats and arrangement of data fields and field prompts stem from Smith and Mosier (1986). Rules for interaction object selection and layout have been extracted from the CUA (IBM, 1992), OSF/Motif (OSF, 1994) and Open Look (Sun, 1990) style guides with regard to national and international standards ISO (1994) and DIN (1988).

## 4.2 Parameter Determination

The second generation step is used to determine the attribute values for the selected abstract interaction objects. Attributes which only depend on the contents of the application (i.e. text labels, field length, format and default values) are copied from the property sheets. A second type of attributes is obtained by the default values which exist for each abstract interaction object type. Default values can be specified by the users of GENIUS in order to define their own style. The third type of attributes, mainly geometric and layout information, depends on other interdependent interaction objects. In this step, the size of each object and the relative position of the objects inside a group are calculated.

## 4.3 Arrangement of Interaction Objects

The layout is determined in the third and last step of the generation. Here, the arrangement within complex elements such as groups is determined first. Afterwards the remaining elements are arranged on the available screen space according to their priority or the given sequence.

The layout is carried out by application of layout rules. These layout rules are defined with regard to the application-specific structure of the data and the laws of perception and cognition for the human visual system such as the law of proximity, similarity, consistency, continuity and symmetry.

## 4.4 Transforming the Generated User Interface Description into a UIMS Specification

The generation process renders an independent user interface description which is transformed into a specification interpreted by the underlying user interface management system (UIMS). This transformation step enables the use of various user interface management systems in combination with GENIUS and allows to support different environments.

## 5 GENERATION OF A USER INTERFACE FOR A PRODUCTION PLANNING AND CONTROL SYSTEM

During the last few years, computer supported systems have been increasingly used for production planning and control. In the beginning, production planning and control systems were limited to mainframes and midrange computers. Therefore these systems have been beyond the means of small and medium sized enterprises. In the last years the situation has changed completely because of the dissemination of powerful personal computers and Unix workstations. As a consequence, implementors of production planning and control systems have been forced to transfer their systems to new hardware platforms and operating systems. The widespread use of window systems on these platforms such as the X Windows System for Unix, Presentation Manager for OS/2 or MS-Windows for DOS leads not only to the transfer of the systems onto new platforms, but also to the creation of graphical user interfaces. Most of the implementors have had to migrate the alphanumerical interfaces of their systems to graphical user interfaces. AS/400 applications for instance are provided with an OS/2 front-end with Presentation Manager, MS-DOS applications receive a Windows user interface and HP 3000 or VMS applications are transferred to Unix and receive a OSF/Motif interface. This migration does not only imply the new layout of old screens in form of windows and the use of graphical user interface elements, but also the complete redesign of the dialogue. This redesign makes it possible to consider software ergonomic design guidelines and thus to ensure more efficient operation sequences. For the implementor of production planning and control systems the following challenges arise:

- Building up knowledge for the design and implementation of graphical user interfaces.
- High expenditure of work for the creation of graphical user interfaces.
- Short development cycles in order to bring out early the new product.
- New design of the dialogue.
- Design of graphical user interface.
- Consideration of software ergonomic guidelines and style guides.

This background given, GENIUS has been used in different projects for creating graphical user interfaces for operational information systems. On the one hand projects have been concerned with developing new production planning and control systems for Unix platforms under OSF/Motif; other projects deal with the migration of MS-DOS applications to MS-Windows.

By the migration of alphanumerical interfaces to graphical ones the first design could be received very quickly when using GENIUS. The automatic generation produces a first layout of windows and of the dialogue while considering basic software ergonomic design guidelines. As a starting point the data model have been used. The migration made it possible to refer to an existing data model or to derive the data model from the existing system. An existing data model is referred to when the systems uses a relational data base management system.

Central to the design of new systems is the determination of the requirements for system design. In these cases, GENIUS was used to generate prototypes. These prototypes served as the basis for discussions between users, customers and developers. The results received from the evaluation of the prototypes have been used to complete the data model. The developed data model and the determined functions have been used for generating a first sketch of the final user interface.

In the following sections the individual steps for generating a graphical user interface for a small, customer-oriented part of a production planning and control system will be shown .

## 5.1 Definition and Generation of the System Entry

The data and their relationships are described in an entity relationship model. This model forms the conceptual data model for the application. For definition the system entry, a coarse, not normalized data model without attributes is used. With a direct manipulation editor the elements for the system entry are selected.
The generated window for the system entry shows the basic customer-oriented areas of a production planning and control system (Figure 5). This comprises the components customer, orders, articles, workstations and resources as well as disposition, purchase, sales, store and manufacturing.

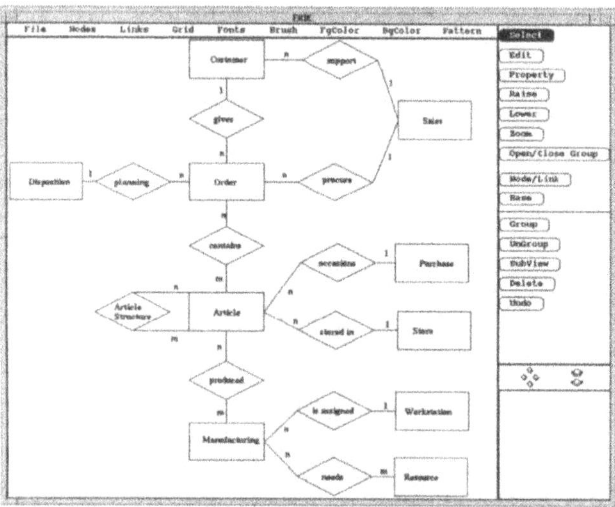


**Figure 4** Coarse data model for a production planning and control system in form of an entity relationship model.

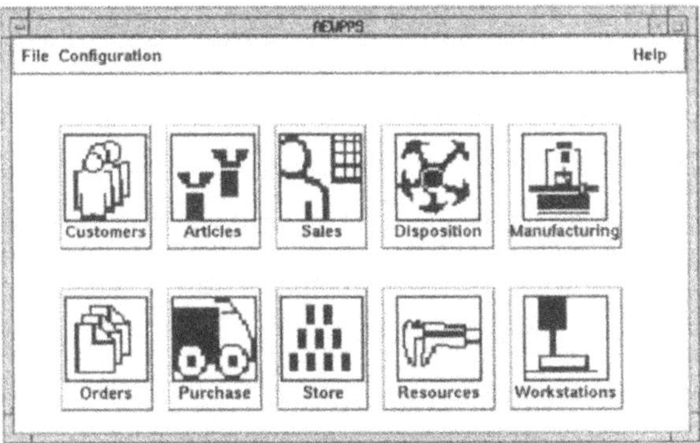


**Figure 5** Generated window of the system entry.

## 5.2 Definition and Generation of Selection Windows

In order to provide fast access to the data, two different access paths are provided. Depending on the number of data records which are available for an object a list or a selection window is generated.

If only a few data records are available for an object, the complete data will be shown in a list. It is possible to manipulate the data shown in the list. Therefore appropriate functions are defined and presented in a menu bar.

For many data records the list will not be clear. In this case the data elements which can be used to identify the desired data are shown in a dialogue box (Figure 6). The functions which are defined for the selection and shown in the dialogue box as push buttons are select, reset, cancel and help. The results of the selection process are presented in a list which the same structure as the selection list for few data records.

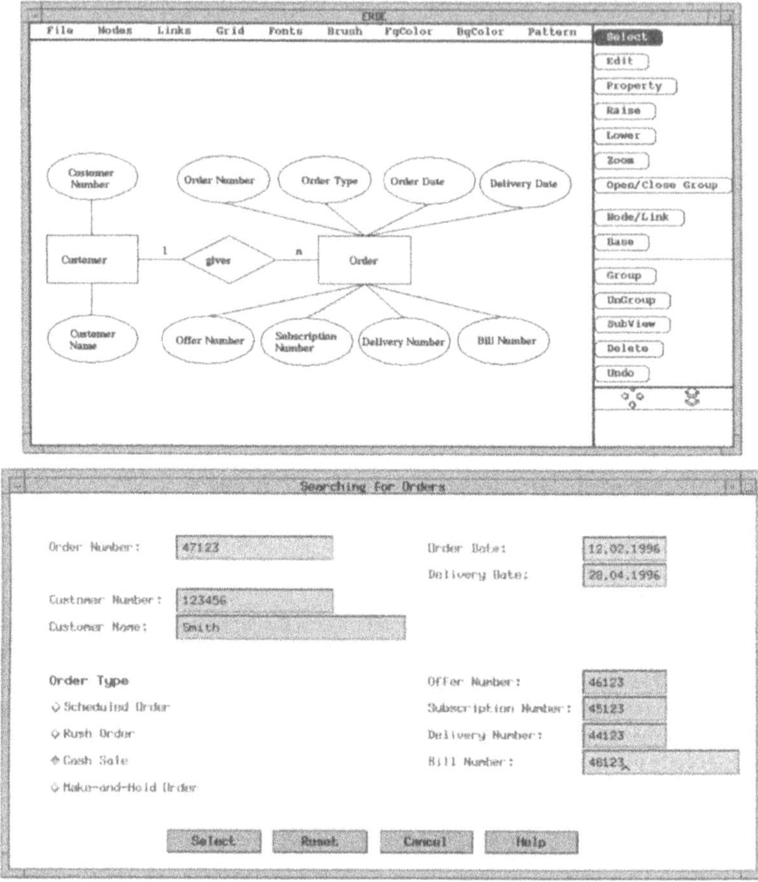


**Figure 6** View definition for a selection and the generated window.

## 5.3 Definition and Generation of a Detail Window

In order to manipulate single data elements of a specific object, a detailed view is defined and the appropriate functions are assigned. Figure 7 shows the defined view for an order and the order positions as well as the generated window.

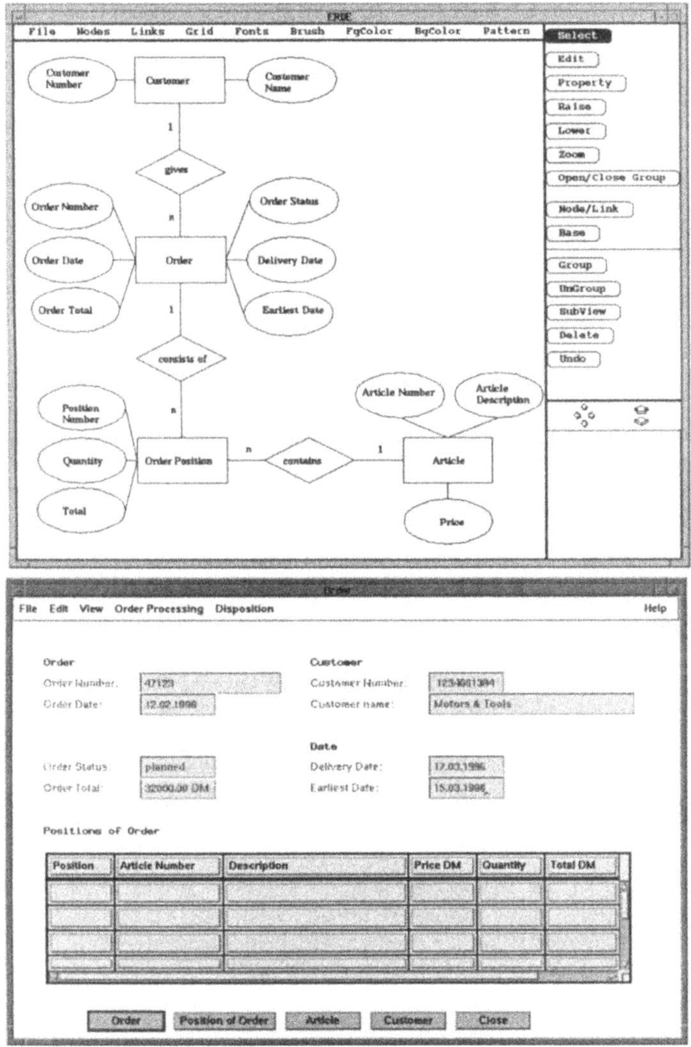


**Figure 7** View definition of an order and the generated window.

## 6 CONCLUSION

If compared to the traditional realisation of graphical user interfaces with user interface management systems, the automatic generation from the data model significantly reduces the effort for application development. In addition, the generation guarantees consistency of the user interface within an application and across different applications.

The automatic generation of user interfaces from data models is not limited to the presentation of information in single windows. It also produces entire dialogue sequences for data-oriented applications. This reduces the effort and simplifies the design of user interfaces. Moreover, such dialogue sequences support the standardization of the dynamic part of the user interface which is not yet covered by existing style guides nor by user interface tools.

With regard to the dissemination of object oriented concepts in the industrial area during the last few years, the data model which is used as a starting point for the generation will be replaced by an object model. In contrast to the data model, an object model has the advantage, that it treats data and functions as a whole which simplifies the assignment of functions to views.

GENIUS provides an integration of user interface development with general software engineering methods. In the early development phases, the automatically generated versions of the user interface support the combination of traditional software engineering methods with prototyping. Such an integration was already demanded by Floyd (1988) and is now widely used in industrial software development projects (Kieback et al., 1992). The automatic generation does not only support the easy and fast production of prototypes but also provides the direct relationship between prototype and specification. The prototypes are the basis for an early design evaluation by the user. This evaluation can provide essential feedback for the specification of the entire software system, and can contribute to ensure the correct realization of the user requirements.

## 7 REFERENCES

Bittner, U.; Hesse, W.; Schnath, J. (1992) Untersuchungen zum Methodeneinsatz in Software-Entwicklungsprojekten. Softwaretechnik-Trends, Band 12, Heft 3, August 1992, 48-60.

Chen, P. (1976) The Entity-Relationship Model - Toward a Unified View of Data. ACM Transactions on Database Systems, Vol. 1, No. 1, 1976, 9-36.

De Baar, D.J.M.J., Foley, J., Mullet, K.E. (1992) Coupling Application Design and User Interface Design, in Proceedings of Human Factors in Computing Systems, CHI '92, May 1992, ACM, New York, 259-266.

DIN-Norm 66234 (1988) Teil 1-9, Bildschirmarbeitsplätze, Normenausschuß Informationsverarbeitungssysteme im DIN. Beuth Verlag, Berlin.

Floyd, Chr. (1984) A Systematic Look at Prototyping, in Approaches to Prototyping (ed. Budde, R.; Kulenkamp, K.; Mathiassen, L.; Züllighoven, H.), Springer, Heidelberg.

Foley; J. D. (1991) User Interface Software Tools, in Telekommunikation und multimediale Anwendungen der Informatik (ed. Encarnacao, J.), GI-21. Jahrestagung, Darmstadt, Springer, Heidelberg.

Foley, J. D.; Kim, W. C.; Kovacevic, S.; Murray, K. (1991) UIDE - An intelligent User Interface Design Environment, in Intelligent User Interfaces (ed. Sullivan, J. W.; Tyler, S. W.), Addison Wesley, Reading, Massachusetts, 339-384.

Foley, J. D.; Kim, W. C. (1990) DON: User Interface Presentation Design Assistant, in Proceedings of the ACM SIGGRAPH on User Interfaces Software and Technology, UIST '90, October 1990, ACM, New York, 10-20.

Hudson, S. E.; King, R. (1986) A Generator of Direct Manipulation Office Systems. ACM Transactions on Office Information Systems 4 (2), 132-163.

Hüttner, J.; Wandke, H.; Beimel, J. (1992) What do system designer know about software ergonomics and how to improve their knowledge?, in WWDU '92 (Work With Display Units), (ed. Luczak, H.; Çakir, A. E.; Çakir, G.), Conference Proceedings, Berlin, E18-E19.

IBM Corporation (1992) Object-Oriented Interface Design: IBM Common User Access Guidelines. Que Corporation, Carmel, IN.

ISO (1994) ISO/WD 9241 Ergonomic requirements for office work with visual display terminals (VDTs), Draft. International Standard Organization.

Kieback, A.; Lichter, H.; Schneider-Hufschmidt, M.; Züllighoven, H. (1992) Prototyping in industrial software projects: Experiences and assessment. Information Technology & People, Vol. 6, No. 2+3, December 1992, 109-143.

Microsoft Corporation (1992) The Windows Interface - An Application Design Guide. Microsoft Press, Redmond.

Myers, B.; vander Zanden, B. (1990) Automatic, Look-and-Feel Independent Dialog Creation for Graphical User Interfaces, in Proceedings of Human Factors in Computing Systems, CHI '90, April 1990, ACM, New York, 27-34.

Olsen, D. R. (1989) A Programming Language Basis for User Interface Management, in Proceedings of Human Factors in Computing Systems, CHI '89, April 1989, ACM, New York, 171-176.

Open Software Foundation (OSF) (1994) OSF/Motif Style Guide, Revision 2.0. Prentice Hall, Englewood Cliffs, N. J.

Petoud, I., Pigneur, Y. (1990) An Automatic and Visual Approach for User Interface Design, in Proceedings of the IFIP TC 2/WG 2.7 Working Conference on Engineering for Human-Computer Interaction (ed. Cockton, G.), North-Holland, Amsterdam.

Singh, G.; Green, M. (1989) Chisel: A system for Creating Highly Interactive Screen Layouts, in Proceedings of the ACM SIGGRAPH Symposium on User Interface Software and Technology, UIST '89, November 1989, ACM, New York, 86-94.

Singh, G.; Green, M. (1991) Automating the Lexical and Syntactic Design of Graphical User Interfaces: The UofA* UIMS. ACM Transactions on Graphics, Vol. 10, No. 3, July 1991, 213-254.

Smith, S. L.; Mosier, J. N. (1986) Guidelines for Designing User Interface Software. Mitre Corporation.

Sun Microsystems (1990) Open Look - Graphical User Interface Application Style Guidelines. Addison-Wesley, Reading, Massachusetts.

Tetzlaff, L.; Schwarz, D. R. (1991) The Use of Guidelines in Interface Design, in Proceedings of Human Factors in Computing Systems, CHI '91, April 1991, ACM, New York, 329-333.

VDMA (1992) Markt für Informationstechnik in der Fertigung wächst weiter. Computerwoche Nr. 41, 09.10.1992.

Weisbecker, A. (1995) A method for the automatic generation of software ergonomic designed user interfaces. Ph.D. Thesis, University of Stuttgart. Springer, Berlin, Heidelberg (in German).

Wiecha, C.; Bennett, W.; Boies, S., Gould, J.; Greene, S. (1990) ITS: A Tool for Rapidly Developing Interactive Applications. ACM Transactions on Information Systems, Vol. 8, No. 3, July 1990, 204-236.

Yourdon, E. (1989). Modern Structured Analysis. Prentice Hall, Englewood Cliffs, N. J.

## 8 BIOGRAPHY

**Prof. Dr.-Ing. habil. Prof. e.h. Dr. h. c. Hans-Jörg Bullinger**

Dr. Bullinger is the head of the Institute for Human Factors and Technology Management (IAT) and the Fraunhofer-Institute for Industrial Engineering (IAO).

In 1978, Dr. Bullinger received the Kienzle-Medal as an award from the University Group of Manufacturing. In 1982 the gold Ring-of-Honour was awarded by the German Society of Engineers (VDI). In 1986 he received the Distinguished Foreign Colleague award from the Human Factor Society. In 1991 he became Honorary Doctor at the University of Novi Sad and Honorary Professor at the University of Science and Technology of China in Hefei. Since 1993 he has been a member of the World Academy of Productivity Science and in 1994 he became an elected honorary member of the Rumanian Society of Mechanical Engineers. In 1995 he received the Arthur Burckhardt Award.

**Dr.-Ing. Dipl.-Math. Klaus-Peter Fähnrich**

Dr.-Ing. Klaus-Peter Fähnrich is currently serving Fraunhofer Society Institute for Industrial Engineering as Deputy Head of Division Information Management. He gained first education in mathematics and computer science. He received a PhD in engineering and is a lecturer at University of Stuttgart.

Currently he is Head of the Departments for Software Management and for Information Systems. His areas of work are Software Management and Software Engineering, Service Engineering, Quality Management and Organisation on Human Factors. He specializes in several branches of industry like machine tool industry, car industry, service industries and media industries.

**Dr.-Ing. Dipl.-Inform. Anette Weisbecker**

Dr.-Ing. Anette Weisbecker studied computer science at the Technical University of Darmstadt and received a PhD in Manufacturing at the University of Stuttgart. She works at the Fraunhofer-Institute for Industrial Engineering (IAO) and currently heads the Group for Software Production. Her area of work is Software Engineering, User Interface Design, Human Factors, with main emphasis on the integration of software engineering methods and user interface design techniques.

## INDEX OF CONTRIBUTORS

# KEYWORD INDEX

GPSR Compliance
The European Union's (EU) General Product Safety Regulation (GPSR) is a set of rules that requires consumer products to be safe and our obligations to ensure this.

If you have any concerns about our products, you can contact us on

ProductSafety@springernature.com

In case Publisher is established outside the EU, the EU authorized representative is:

Springer Nature Customer Service Center GmbH
Europaplatz 3
69115 Heidelberg, Germany

www.ingramcontent.com/pod-product-compliance
Ingram Content Group UK Ltd.
Pitfield, Milton Keynes, MK11 3LW, UK
UKHW012214240726
13966UKWH00002B/753
* 9 7 8 1 4 7 5 7 6 5 3 9 7 *